Mathematics Made Easy @ your Speed

Dr. Jambulingam Subramani

Published by

BONFRING®
Intellectual Integrity

Mathematics Made Easy @ your Speed

First Print: February 2016

ISBN 978-93-85477-70-6

For Further Enquiries and Comments Please Write to

Dr. Jambulingam Subramani, Ph.D.

Plot 38, Mangammal Salai,

Renga Nagar, K.K.Nagar Post

Tiruchirapalli- 620021, Tamilnadu, India

Email:drjsubramani@yahoo.co.in | Mobile: +91-9443753853

Dedicated to My Parents

Thiru Jambulingam Muthuraja

Tmt. Sinnadhal Jambulingam

Preface

People do many sort of mathematical computations invariably from all walks of life. For example students do some sort of computations during class work, while getting bus tickets and purchasing some stationery etc. Similarly vegetable vendors, bank employees, salesmen of departmental stores or of medical shops, conductors of town buses, railway employees are some of the people using mathematical computations for the purchases made by the customers and/or passengers or to pay the balance amounts to the customers and/or passengers. In all such situations the usual calculators or computers cannot be used for effecting the computations due to cost and time factor. So the author has planned to write this book to overcome these difficulties. In this book some new methods of multiplications practised by the author for more than two decades are presented and are explained with the help of numerical examples. Consequently, it is believed by the author that these methods are very much helpful in improving the skills of the readers in numerical computations. Further critical comments and suggestions are invited from the potential readers to improve the contents as well as the presentation of this book.

Dr. Jambulingam Subramani

Acknowledgement

I am extremely grateful to Dr.B.Ilango, Formerly Vice-Chancellor, Bharathiyar University, Coimbatore for his interests and guidance shown on me to undertake this project. I am sure without his guidance and advice I could not have completed this project. I am also very much grateful to Prof.K.N.Ponnuswamy, Prof.C.R.Rao, Prof.M.L.Aggarwal, Prof.M.M.Ramaswamy, Prof.B.Thangaraju, Prof.R.Ethirajulu, Prof.G.Stephen Vincent, Prof.C.Raju, Prof.K.Govindaraju, Prof.Rahul Mukherjee, Prof.A.D.Das, Prof.D.K.Ghosh, Prof.M.R.Srinivasan, Prof.S.Balamurali, Dr.S.DuraiRaju, Mr.R.Seshagiri, Mr.D.Rajakumar, Mr.T.Rajkumar, Mr.Navilu M.Subramaniam and Mr.N.P.Sukumar, who have inspired me a lot in my attitudes and behaviours and to undertake this project. Further I wish to acknowledge my gratitude to my parents Thiru.K. Jambulingam Muthuraja and Tmt.J.Chinnathal, my wife Tmt.S.Bhuvaneshwari, my daughter S.Karthika and my son Mr.S.Karthikeyan for their patience and their continuous support rendered to me for completing this project. Last but not the least I wish to acknowledge all my friends who have helped me in completing this project. Finally I wish to acknowledge those readers in advance who will pass on the critical comments, which will certainly be helpful to improve the contents as well as the presentation in the future editions.

Dr. Jambulingam Subramani

Author's Profile

Dr. Jambulingam Subramani, Ph.D.

Associate Professor

Department of Statistics

Pondicherry University

Puducherry

India

Dr.J. Subramani, a renowned statistician from India currently working as Associate Professor in the Department of Statistics, Pondicherry University, has been working for improving the analytical skills of young students and eliminate math phobia among them for more than three decades by organizing training classes and math exhibitions.

A certified ISO 9000-2000 QMS Lead Auditor, he carries over 25 years of teaching and research experience combined with industry experience. He has also served as visiting scientist in the prestigious University of Windsor, Canada.

His most significant contributions have been in the fields of estimation of variance components particularly the C R Rao's MINQUE (minimum norm quadratic unbiased estimation) theory, design of experiments, sampling theory, Statistical Quality Control and recreational mathematics.

Dr. Subramani, who has authored about 120 research papers, continues to be a peer reviewer for more than 20 highly reputed statistical journals from India and abroad.

Dr. Subramani also had received the young statistician award from the International Statistical Institute, Netherlands, as well as Indian Society of Probability and Statistics. Besides these Awards he has received the "Best Academic Researcher Award" by the Association of Scientists, Developers and Faculties (ASDF), Pearson Teaching Awards- Best Teacher (Male)-Third Edition and Pearson Teaching Awards- Innovations in Teaching in Higher Education- UG- Government- Third Edition, and Outstanding Faculty Award from Venus International Foundation.

CHAPTER 1

INTRODUCTION

In day-to-day operations everyone in the society has to do some sort of mathematical computations due to the nature of their work. In fact it is necessary for people from school students to research scientists and ordinary laymen to top level businessmen for the purpose of computing the school exercises; establishing research findings; calculating their earnings or expenses; assessing their turn over or marketing share, etc.

Without using mathematical computations subjects like science, engineering and technology will serve no purpose to the human society. For example in the booking counters at Bus Terminus, Railway Station, cinema, etc. people come across some sorts of mathematical computations for paying the required money for the purchase of tickets or getting the balance money. Similarly in departmental stores, shops, hotels, town buses and other establishments, it is absolutely necessary to compute the amounts for their purchases where the concerned people have to take immediate decisions to pay the bills. In such situations, it is not always advisable to use calculators and computers to compute the amount to be paid for the bills.

Even sometimes the calculators and computers are not useful in computations, in which the number of integers involved is a very high number. For example, if we assume that every person requires 100 grams of rice per day and the population of Tamilnadu is 61532759, how many grams of rice required for all people of Tamilnadu for 5 years? This requires computing the expression 61532759x100x365x5, which leads to 11229728517500 grams of rice for the people of Tamilnadu for five years.

In such computations the usual or ordinary calculators may not be useful to compute the value, which involves 14 digits. The methods presented in the following sections are aimed to help people to overcome such difficulties and to do the calculations particularly multiplications and squares without the help of calculators and computers. As a result the methods of computations to be discussed in this paper will create awareness and improve the interest in the subject of mathematics among the readers. In this book we are going to discuss several methods of computations with adequate number of numerical examples.

The methods to be discussed in succeeding chapters are: Multiplication by Line Graphs, Cell-wise Multiplication, Modified Cell-wise Multiplication, Multiplication of Numbers Based at 10, Multiplication of Numbers-Based at - 100, Multiplication of Numbers - Based at 1000, Multiplication of Numbers - Based at 10000, Multiplication of Numbers - Based at 50, Multiplication of Numbers-Based at 500, Multiplication of Numbers-Based at 5000, Multiplication of Special Numbers and Squaring of Numbers together with some mathematical puzzles. The author strongly feels that it will certainly help the readers, particularly students who prepare for the aptitude tests for their placements to enhance their numerical ability after reading this book.

CHAPTER 2

MULTIPLICATION BY LINE GRAPHS

In this chapter a new type of multiplication namely Multiplication by Line Graphs has been discussed. The method is called so for it involves only Horizontal lines and Vertical lines.

This procedure is based on the counting principles. It does not require any multiplication or carry over during the multiplications but requires only drawing some Horizontal and Vertical lines as required. Hence this method can also be called as HV Lines multiplication. Here straight lines are used to represent the numbers. For example, the number 1 can be represented by a single straight line; 2 can be represented by two straight lines drawn adjacently; 3 can be represented by three straight lines drawn adjacently and so on. Further thin lines are used to represent the numbers at 1st position and thick lines are used to represent the numbers at the 10th position and so on. That is why the lines used to represent the numbers at the 10th position are thicker than the numbers at 1st position. The thickness of the lines increases as the position of the number increases. Further any multiplication involves two numbers and is respectively represented by Vertical lines and Horizontal lines. Please note that there is no specific criteria to choose the numbers to be represented by vertical lines as well as Horizontal lines due to the property of axb=bxa in multiplication.

The following table gives the notations to be used in the method of multiplication by line graphs:

Table 2.1: Notations to be used in Multiplication by Line Graphs

Numbers	Notations by Vertical lines	Notations by Horizontal lines
1		
2		
3		
12		
32		

This method involves the following steps:

i.   Let m and n be the two single digit numbers to be multiplied. Then draw m horizontal lines and n vertical lines.

ii. Count the number of intersections (points to be met by the Horizontal and Vertical lines), which gives the value of the multiplication mxn.

iii. If the counting exceeds the value 10 reduce the value to mod 10 and carry over the remainder to next position.

To understand the above method of multiplication, consider the examples given below:

Example 2.1: To find the value of 2x3, draw 2 horizontal lines and 3 vertical lines as given in Figure 2.1.

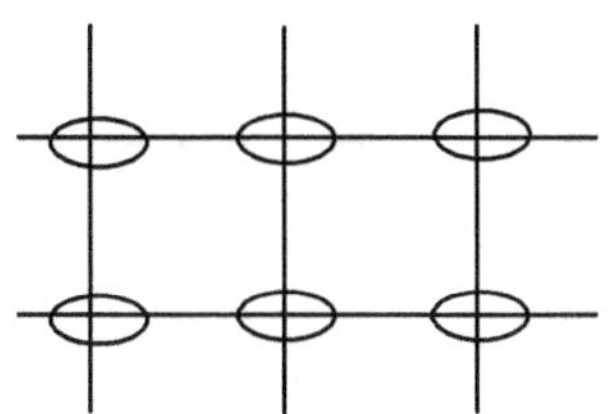

Figure 2.1: 2x3 = 6

By counting the number of intersections, one may get 2x3 = 6. The intersections are marked by ◯

It is not necessary to mark the intersections by a circle, but it is done for the convenience of the readers to understand the concepts clearly; Particularly the marking is more useful when one may come across different type of interactions. For more details refer the examples to be discussed later.

Example 2.2: To find the value of 3x5, draw 3 horizontal lines and 5 vertical lines as given in Figure 2.2

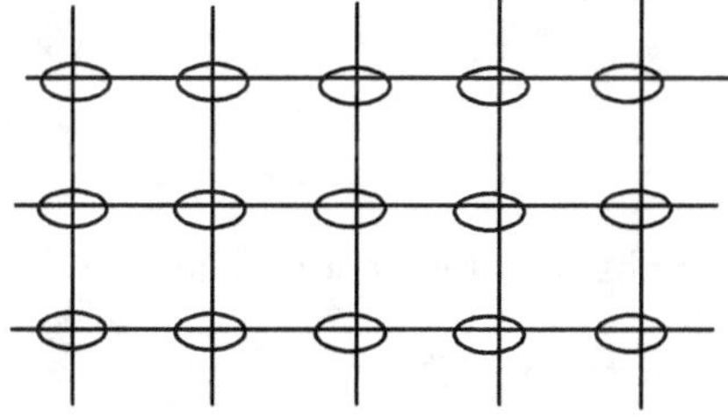

Figure 2.2: 3x5 = 15

By counting the number of intersections, one may get 3x5= 15. The intersections are marked by 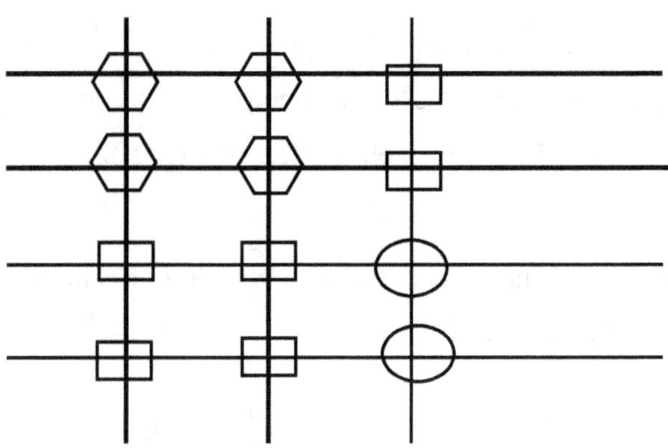

Example 2.3: To find the value of 3x0, draw 3 horizontal lines and 0 vertical lines as given in Figure 2.3.

Figure 2.3: 3x0 = 0

By counting the number of intersections (Since there is no intersection), one may get the value of 3x0=0.

Example 2.4: To find the value of 22x21, draw 2 thick and 2 thin horizontal lines and 2 thick and 1 thin vertical line as given in Figure 2.4.

Figure 2.4: 22x21 = 462

By counting the number of intersections, made by

1.	Thin lines, denoted by

2.	Thin and Thick lines, denoted by

3.	Thick lines, denoted by

and writing them from right to left one may get the product of these numbers. That is, write the number of 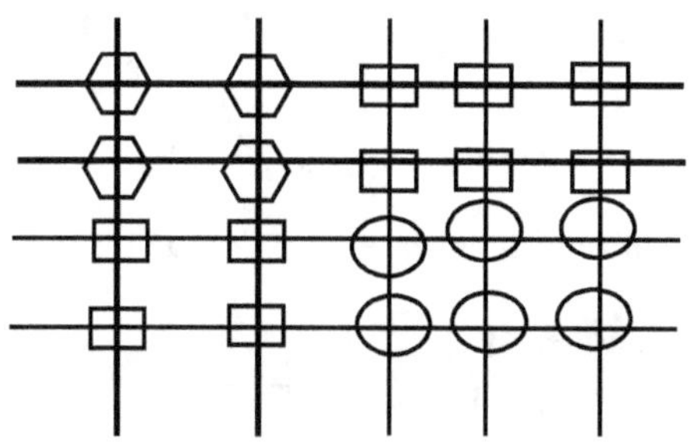 intersections in each of the category mentioned above and in the same order to get the product of two digit numbers. It is to be noted that a thin line can represent the value 1 in the 1st position where as the thick line represent the value 1 in the 10th position. From the Figure 2.4 one can easily obtain that 22x21 = 462.

Example 2.5: To find the value of 22x23, draw 2 thick and 2 thin horizontal lines and 2 thick and 3 thin vertical lines as given in Figure 2.5.

Figure 2.5: 22x23 = 506

By counting the number of intersections, made by

1. Thin lines, denoted by
2. Thin and Thick lines, denoted by
3. Thick lines, denoted by

and writing them from right to left one may get the product of these numbers. That is, ,write the number of intersections in each of the category mentioned above and in the same order to get the product of two digit numbers. It is to be noted that a thin line can represent the value 1 in the 1st position where as the thick line represent the value 1 in the 10th position. From the figure one may get easily that 22x23 = 506.

Example 2.6: To find the value of 13x22, draw 2 thick and 2 thin vertical lines and 1 thick and 3 thin horizontal lines as given in Figure 2.6.

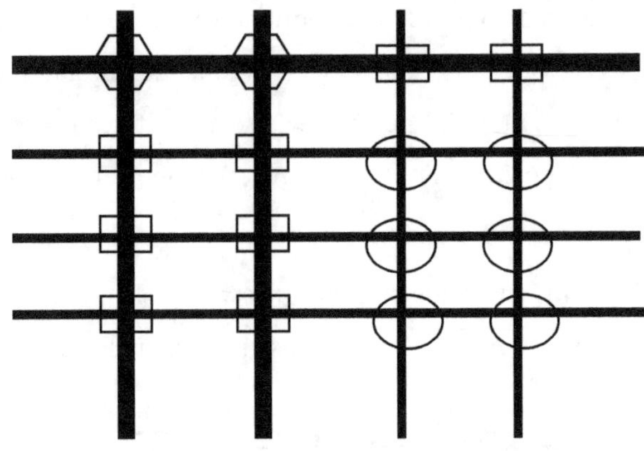

Figure 2.6: 13x22 = 286

By counting the number of intersections, made by

1. Thin lines, denoted by ⬭

2. Thin and Thick lines, denoted by ▢

3. Thick lines, denoted by ⬡

and writing them from right to left one may get the product of these numbers. That is, ╪ ╪ ╪ ,write the number of intersections in each of the category mentioned above and in the same order to get the product of two digit numbers. It is to be noted that a thin line can represent the value 1 in the 1st position where as the thick line represent the value 1 in the 10th position. From the figure one may get easily that 13x22 = 286.

Remark: Alternatively one may use same type (thickness) of lines to represent first & tenth position numbers by leaving some spaces between them. The example given below will explain this method of computation.

Example 2.7: To find the multiplication of 22 and 13 one may use the following steps: Here the same thickness lines are used to represent the first and tenth position numbers but some spaces are left in between to distinguish the numbers. Similarly one can use different coloured lines to represent numbers with different bases.

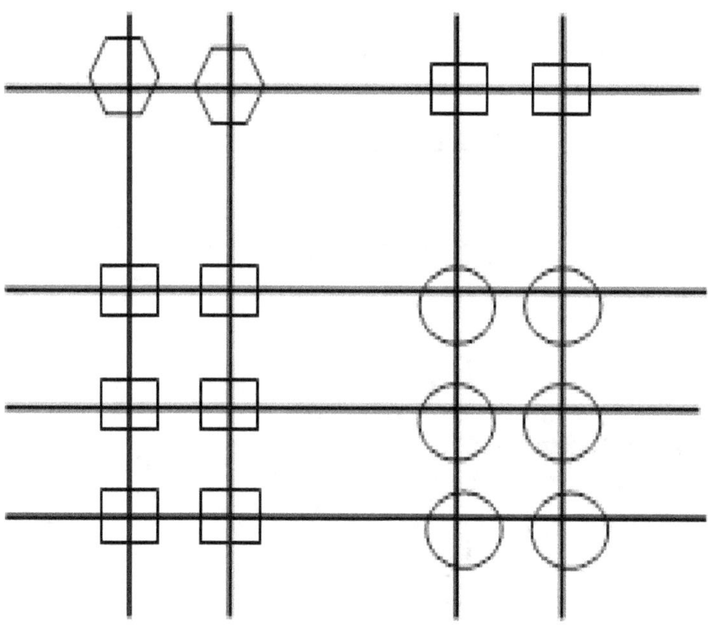

Figure 2.7: 13x22 = 286

By counting the intersections one may get 13x22= 286. That is, by counting the intersections represented by different shapes as discussed earlier.

SOLVED EXAMPLES

The following are some of the solved examples given to explain the method of Multiple by Line Graphs.

Problem 2.1: To find 3x1

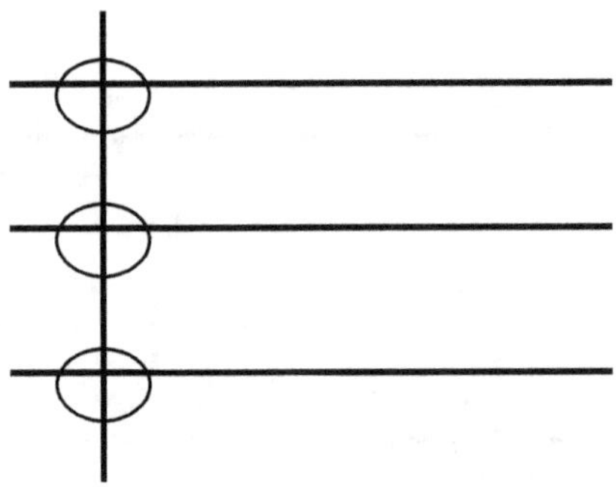

Solution: 3x1=3

Problem 2.2: To find 3x4

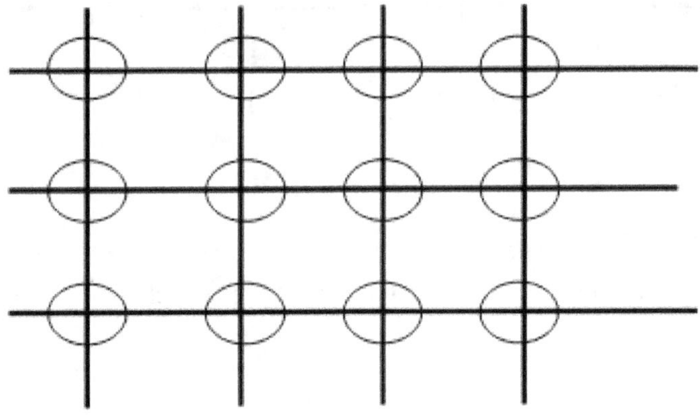

Solution: 3x4=12

Problem 2.3: To find 4x5

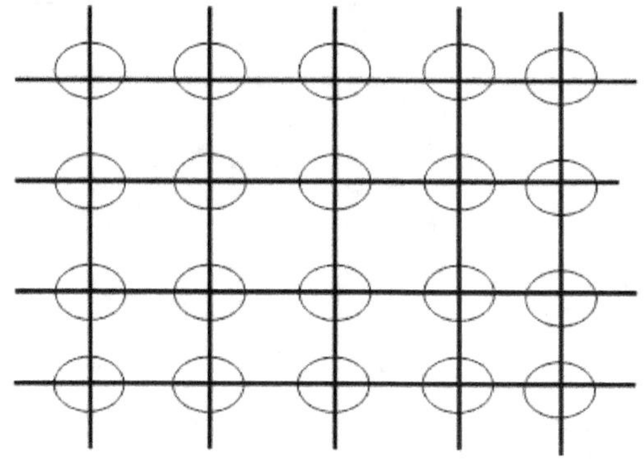

Solution: 4x5=20

Problem 2.4: To find 6x7

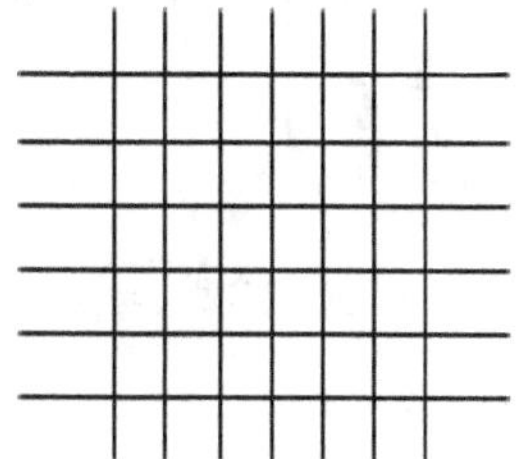

Solution: 6x7=42

Problem 2.5: To find 24x43

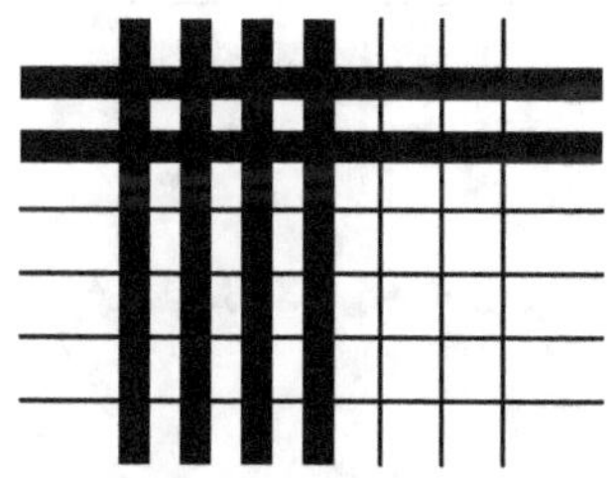

Solution: 24x43=1032

Problem 2.6: To find 6x30

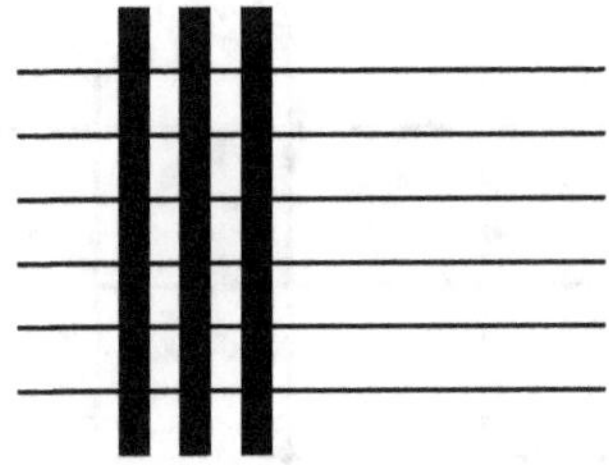

Solution: 6x30=180

Problem 2.7: To find 40x30

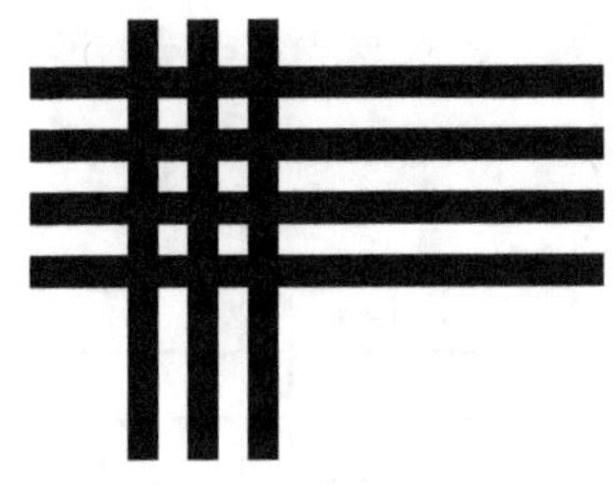

Solution: 40x30=1200

Problem 2.8: To find 13x32

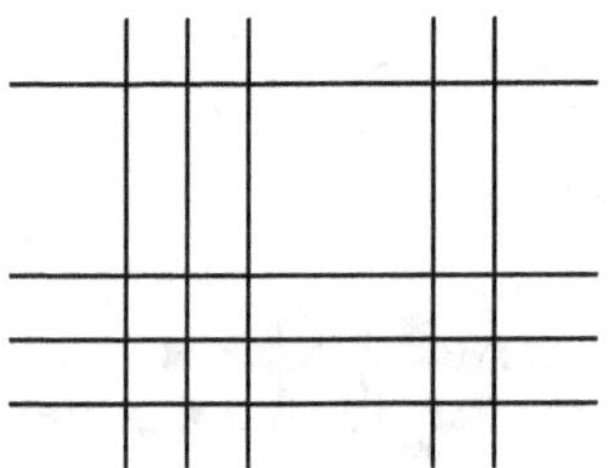

Solution: 13x32=416

Problem 2.9: To find 32x22

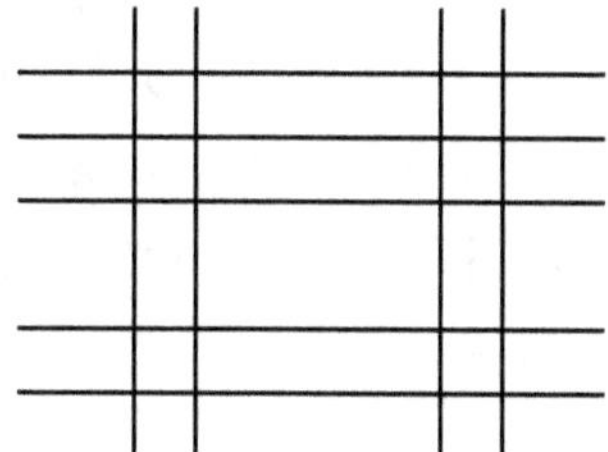

Solution: 32x22=704

Problem 2.10: To find 6x3

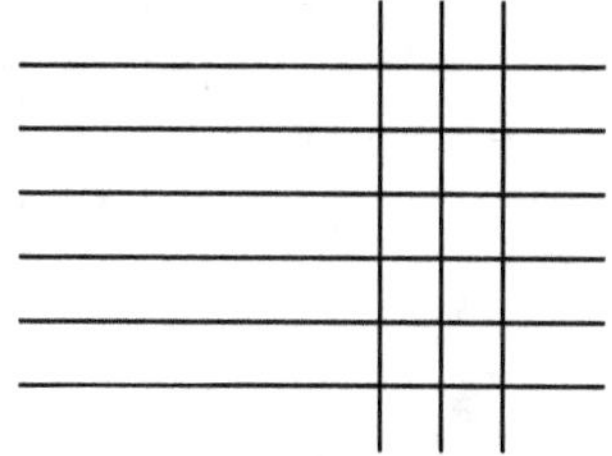

Solution: 6x3=18

Problem 2.11: To find 30x21

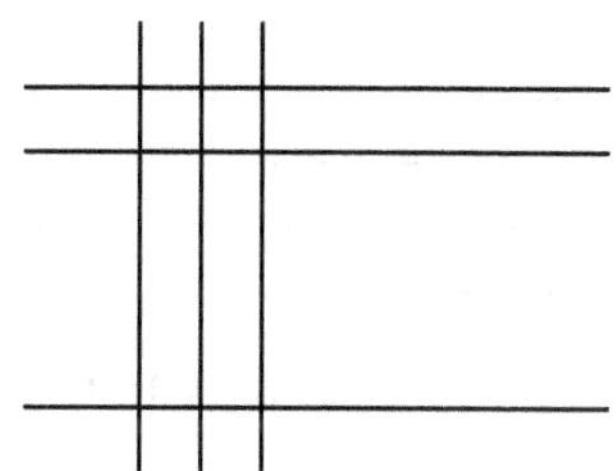

Solution: 21x30=630

EXERCISE PROBLEMS

The following are some of the exercise problems to be solved by using the method of Multiple by Line Graphs to have an acquaintance with the new procedure. Write down the numbers to be multiplied and their results from the following Line Graphs?

Problem 1

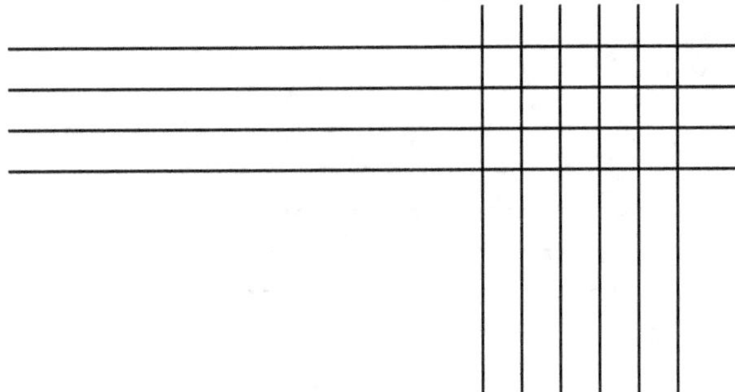

Problem 2

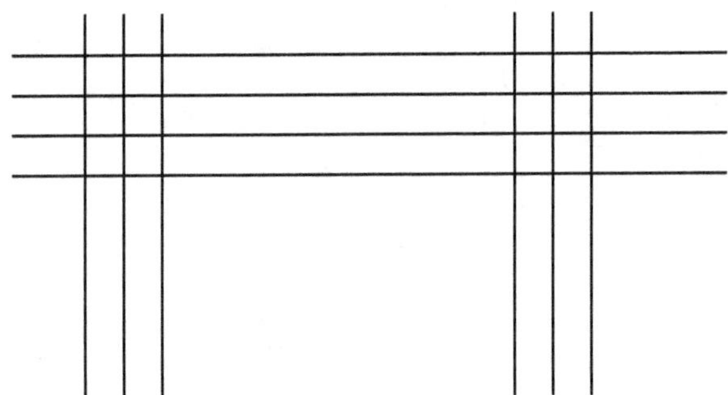

Problem 3

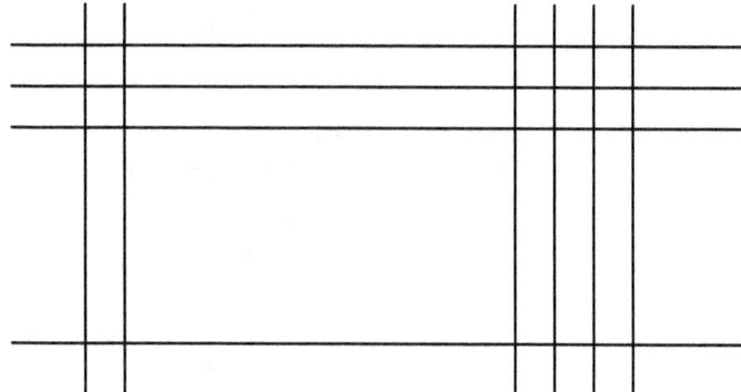

Problem 4

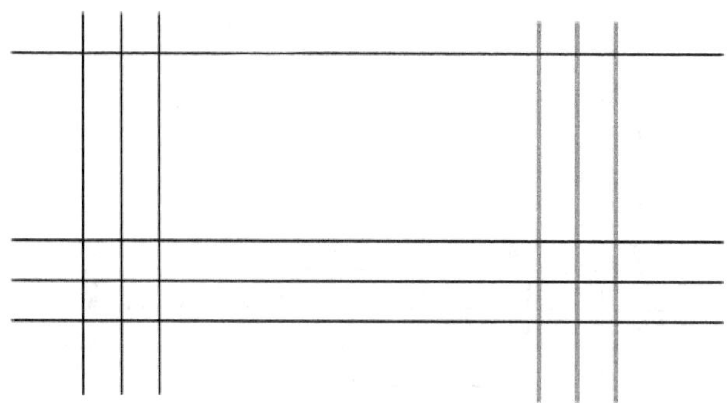

Problem 5

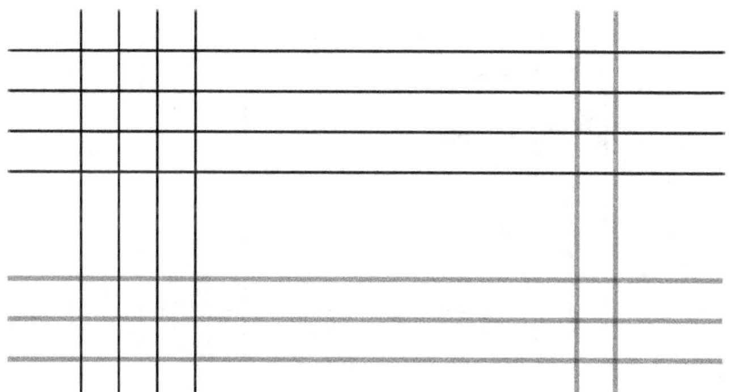

Problem 6

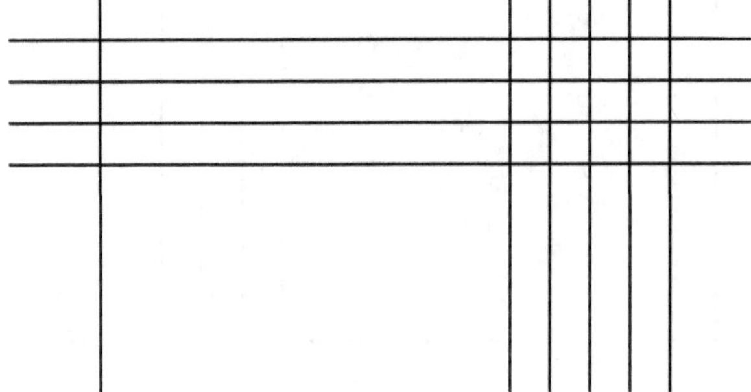

Problem 7

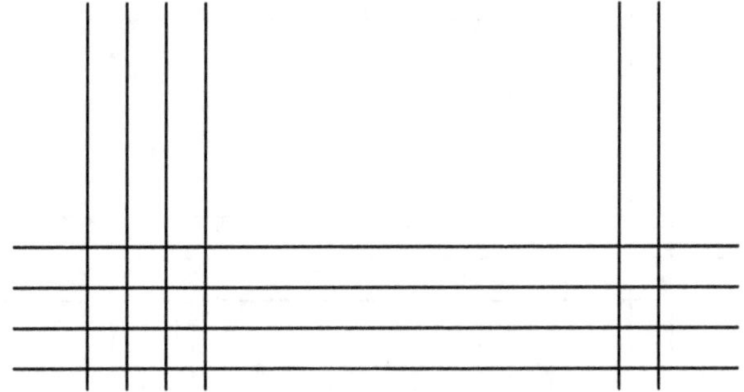

Problem 8

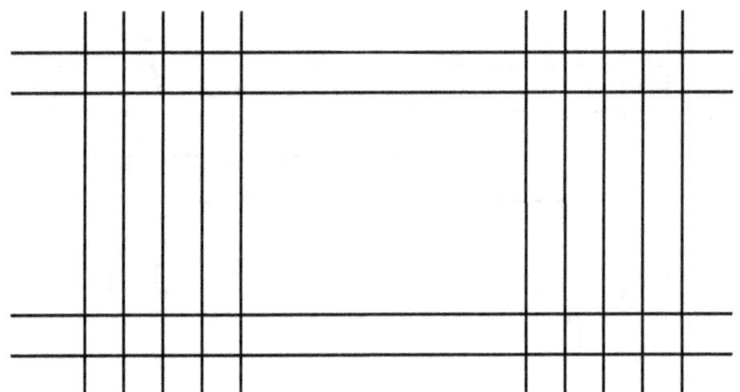

Problem 9

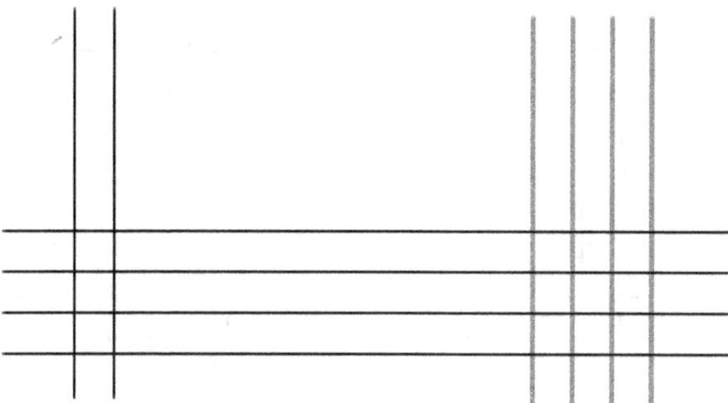

Problem 10

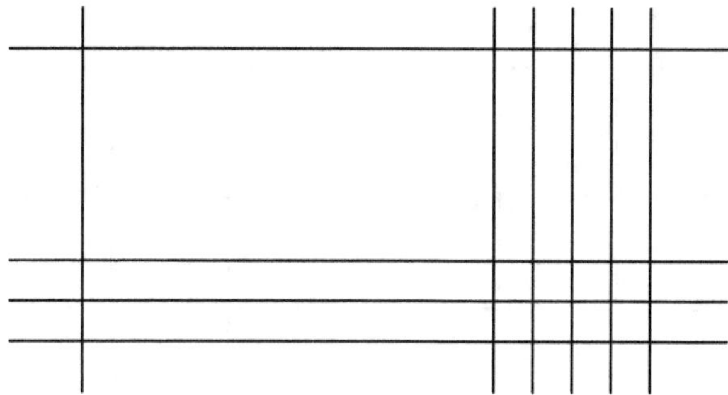

Problem 11

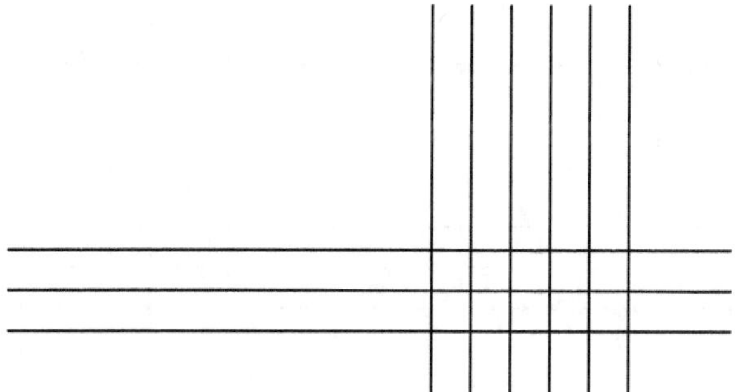

Problem 12

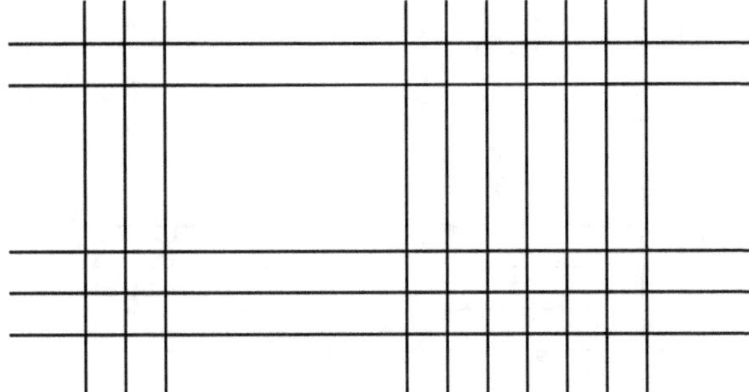

Problem 13

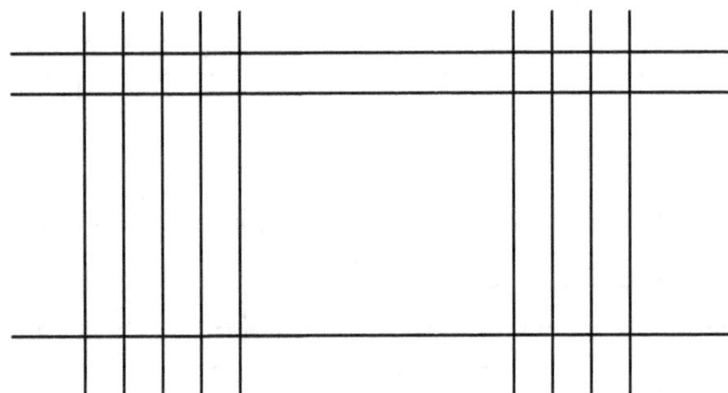

Problem 14

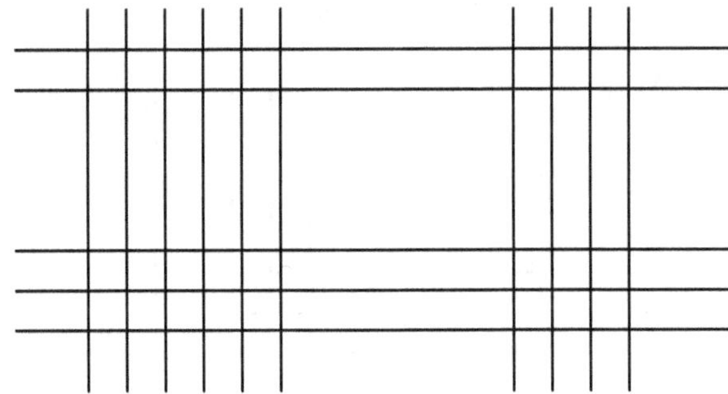

Problem 15

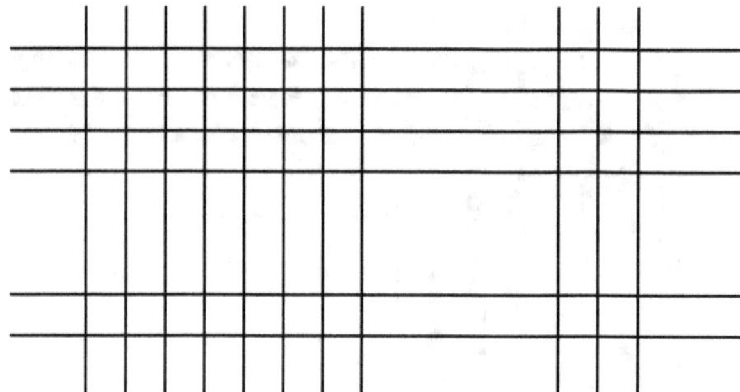

Problem 16

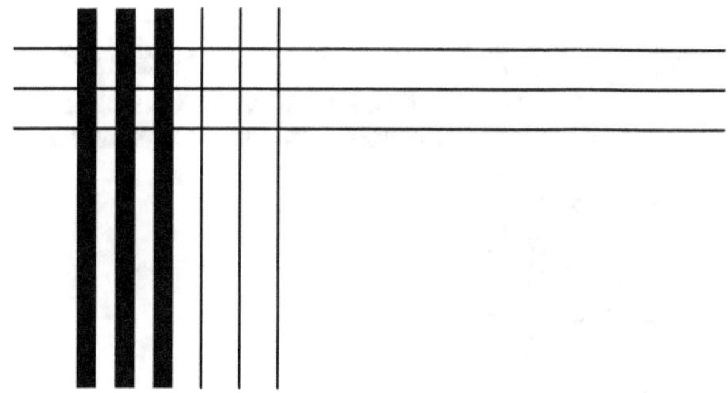

Problem 17

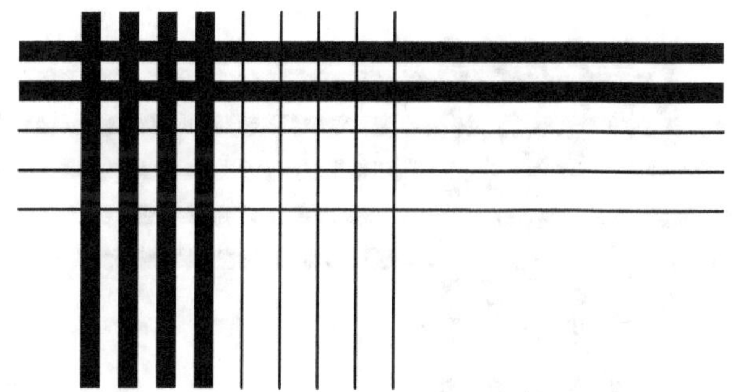

Problem 18

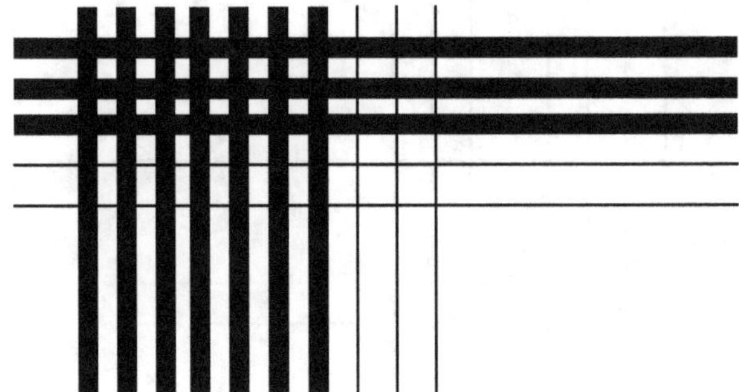

Problem 19

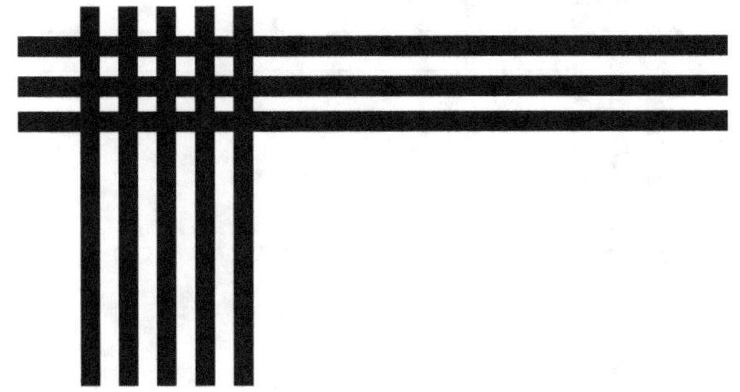

Problem 20

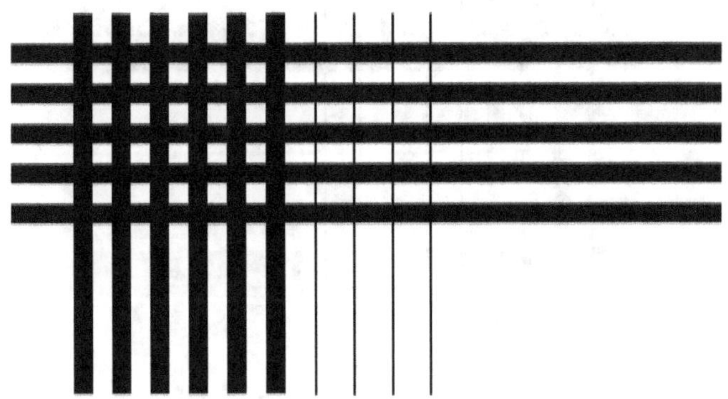

Problem 21

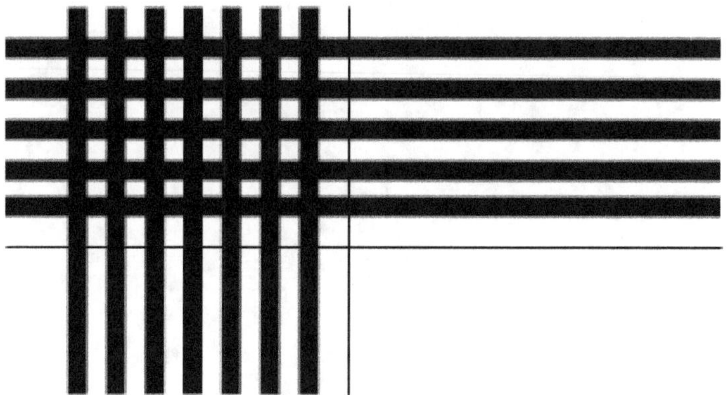

Chapter 3

Cell-Wise Multiplication

In the previous chapter we discussed a new method of multiplication namely "Multiplication by Line Graphs". This method is a graphical procedure for multiplication. Further it involves only horizontal and vertical lines and the counting principle and hence it will attract the youngsters particularly school students. However the problem is to draw a large number of lines whenever we want to multiply two numbers with large number of digits. Hence this procedure is very much useful and interesting to multiply two digit numbers but not simple to multiply more than two digit numbers.

An alternative procedure namely "Cell-wise Multiplication" is to be discussed in this chapter. This procedure is based on the cell-wise multiplication principles. It does not require any carry over during the multiplication but requires only the multiplication of numbers from 1 to 9. Consequently this procedure is very compact and occupies little space; it is easy to diagnose due to cell-wise multiplication; it is easy to multiply any two numbers with any number of digits. It is to be noted that sometimes we cannot compute the correct numbers even using calculators or computers. For example, if we want to multiply the numbers 123456789 and 987654321 we cannot use either calculator or computer as it is. Suppose if we use computer by using some software we can compute the product of the above two numbers. However by using the Cell-wise Multiplication principle, one can obtain the product of two numbers with any number of digits.

The steps involved in this procedure are as follows:

i. Let the number of digits in the given numbers be m and n respectively

ii. Draw mxn array with mn cells.

iii. Write the first number vertically to the left of the array from the top to down such that i^{th} digit from left corresponding to the ith row and is denoted by R_i.

iv. Write the second number horizontally to the top of the array from left to right such that the jth digit from left corresponding to the jth column and is denoted by C_j.

v. Find the product of ith digit of first number and j^{th} digit of the second number and write the value at the $(i,j)^{th}$ cell. That is, the entry at the $(i,j)^{th}$ cell is $R_i x C_j$.

vi. Start from the $(m,n)^{th}$ cell, find the sum of the numbers appeared in the secondary diagonals and write the results from right to left.

vii. If any value is more than 10 reduce to mod 10 and carry over the remainder to the next value.

viii. Continue until to reach the cell (1,1).

For more details one may refer the solved examples given below.

Example 3.1: To find out the product of 12x14, draw an array with 2x2=4 cells as given in the Figure 3.1. Write down the numbers 12 and 14 as written in the figure then find the products of 1x1, 1x4, 2x1 and 2x4 and write the results in their respective cells. Find the sum of the numbers in the secondary diagonals starting from the cell (2,2), one may get 8,6 and 1. Finally writing these values from right to left one may get the value of 12x14 = 168 as given in Figure 3.1.

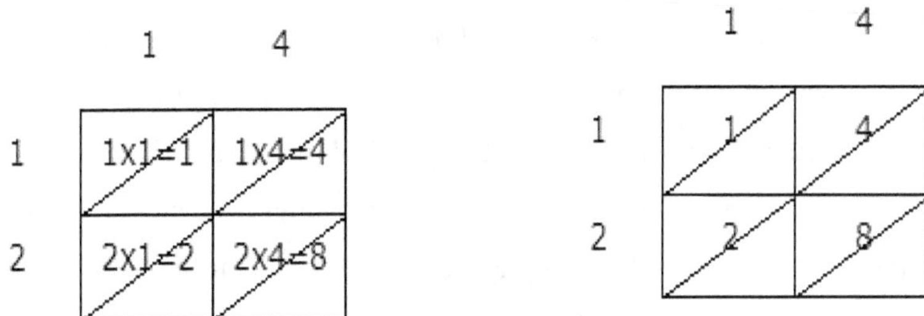

Figure 3.1: 12x14= 168

Example 3.2: To find out the product of 122x142, draw an array with 3x3=9 cells as given in the Figure 3.2. Write down the numbers 122 and 142 as written in the figure then find the products of 1x1, 1x4, 1x2, 2x1, 2x4, 2x2, 2x1, 2x4 and 2x2, and write the results in their respective cells. Find the sum of the numbers in the secondary diagonals starting from the cell (3,3) one may get 4, 2, 3, 7 and 1. Finally writing these values from right to left one may get the value of 122x142 = 17324 as given in Figure 3.2.

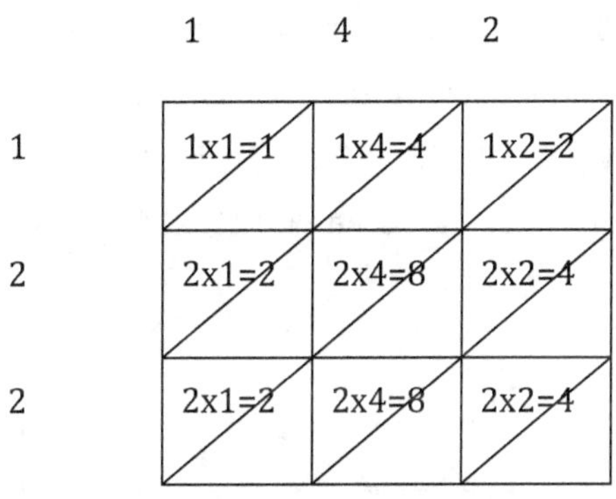

Figure 3.2: 122x142= 17324

Example 3.3: To find out the product of 122x1423, draw an array with 3x4=12 cells as given in the Figure 3.3. Write down the numbers 122 and 1423 as written in the figure then find the products of 1x1, 1x4, 1x2, 1x3, 2x1, 2x4 ... 2x3 and write the results in their respective cells. Find the sum of the numbers in the secondary diagonals starting from the cell (3,4) one may get 6, 0, 6, 3, 7 and 1. Finally writing these values from right to left one may get the value of 122x1423 = 173606 as given in Figure 3.3.

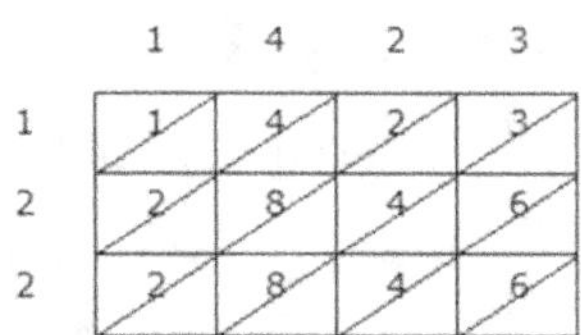

Figure 3.3: 122x1423= 173606

Example 3.4: To find out the product of 1213x1415, draw an array with 4x4=16 cells as given in the Figure 3.4. Write down the numbers 1213 and 1415 as written in the figure then find the products of 1x1, 1x4, 1x1, 1x5, 2x1, 2x4, 2x1, 2x5 an so on and write the results in their respective cells. Find the sum of the numbers in the secondary diagonals starting from the cell (4,4), one may get 5, 9, 3, 6, 1, 7 and 1. Finally writing these values from right to left one may get the value of 1213x1415 = 1716395 as given in Figure 3.4.

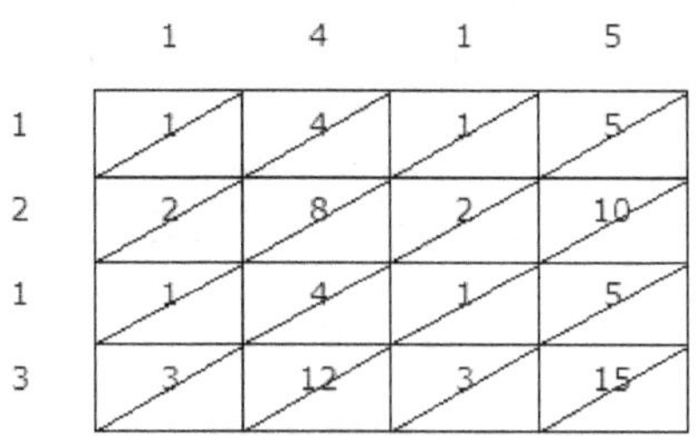

Figure 3.4: 1213x1415= 1716395

Example 3.5: To find out the product of 123456789x1234567890, draw an array with 9x9=81 cells as given in the Figure 3.5. Write down the numbers 123456789 and 123456789 as written in the figure then find the products of 1x1, 1x2, 1x3, 1x4, 1x5, … 9x6, 9x7, 9x8 and 9x9 and write the results in their respective cells. Find the sum of the numbers in the secondary diagonals starting from the cell (9,9) one may get 1,2,5,0,9,1,0,5,7,8,7,5,1,4, 2,5 and 1. Finally writing these values from right to left one may get the value of 123456789x123456789 =15241578750190521 as given in Figure 3.5.

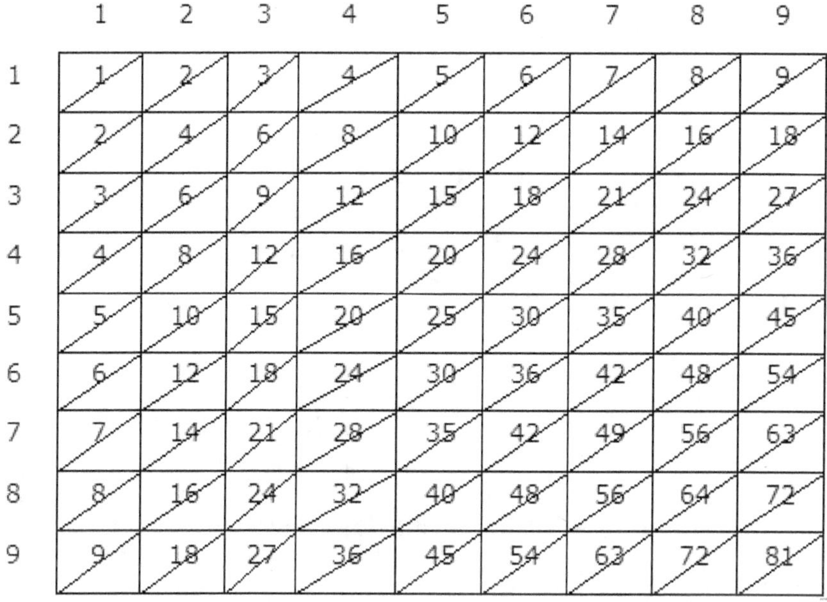

Figure 3.5: 123456789x123456789=15241578750190521

SOLVED PROBLEMS

The following are some of the solved problems given to explain the method of Cell-wise Multiplication:

Problem 3.1: To find 12x13

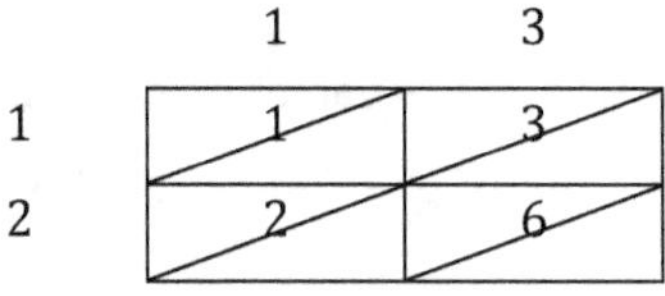

Solution: 12x13=156

Problem 3.2: To find 16x24

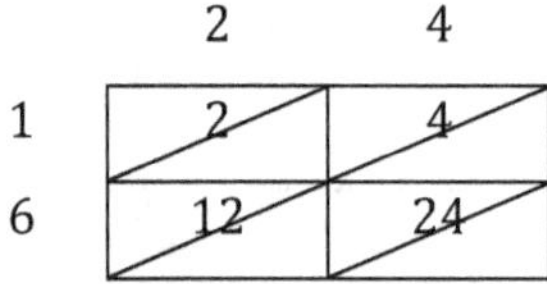

Solution: 16x24=384

Problem 3.3: To find 15x14

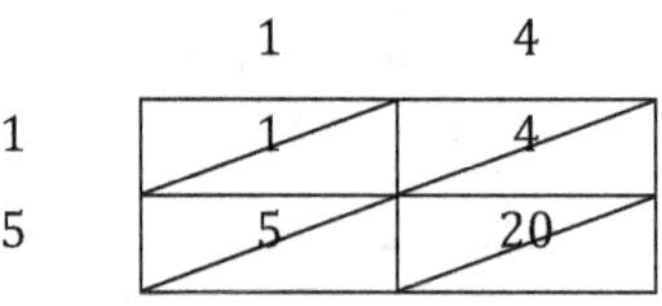

Solution: 15x14=210

Problem 3.4: To find 16x245

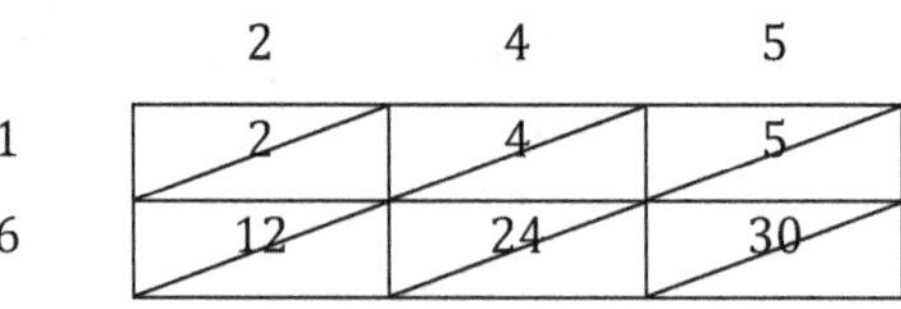

Solution: 16x245=3920

Problem 3.5: To find 123x324

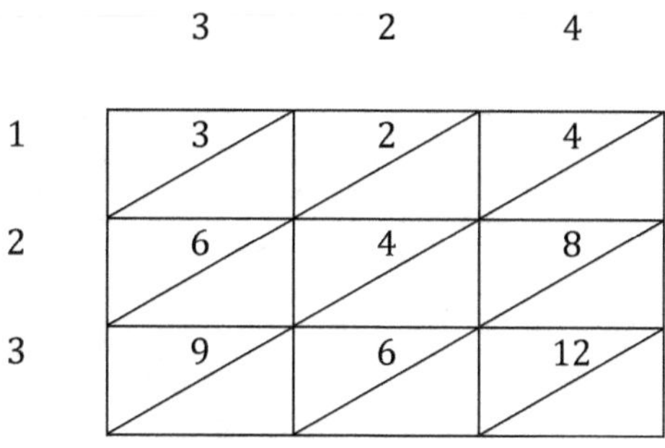

Solution: 123x324=39852

Problem 3.6: To find 223x325

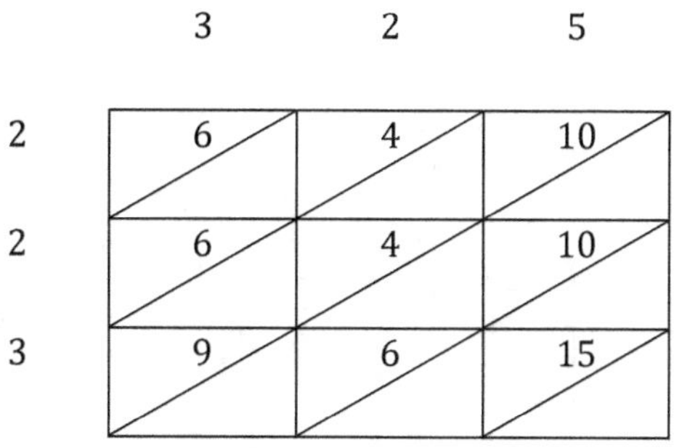

Solution: 223x325= 72475

Problem 3.7: To find 122x122

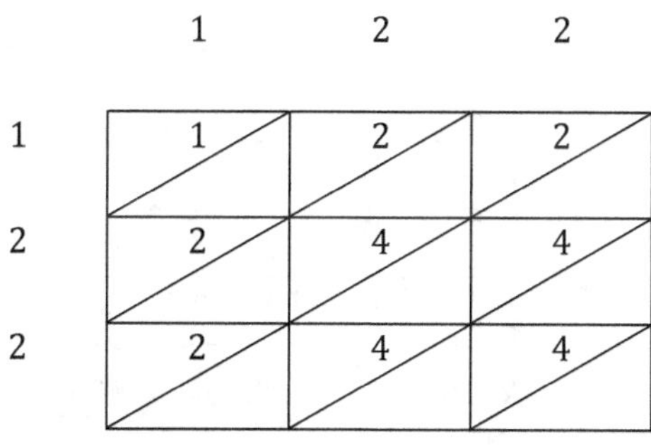

Solution: 122x122 = 14884

Problem 3.8: To find 512x512

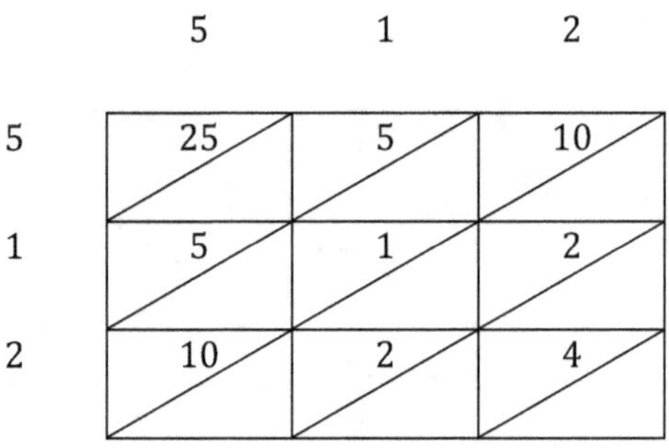

Solution: 512x512 = 262144

Problem 3.9: To find 122x1225

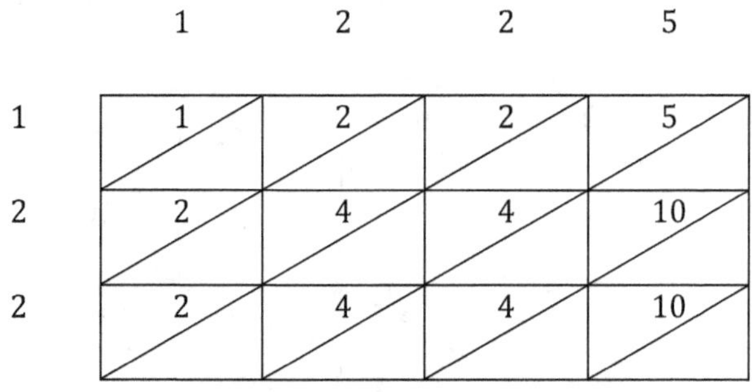

Solution: 122x1225 = 419450

Problem 3.10: To find 1225x1225

Solution: 1225x1225 = 1500625

Problem 3.11: To find 1234x5678

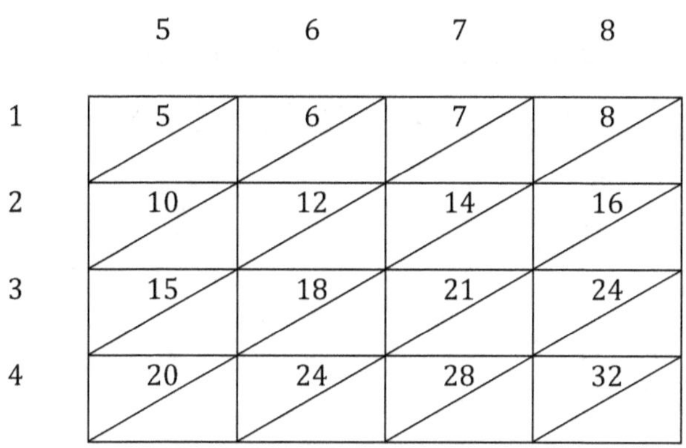

Solution: 1234x5678 = 7006652

Problem 3.12: To find 1234x56789

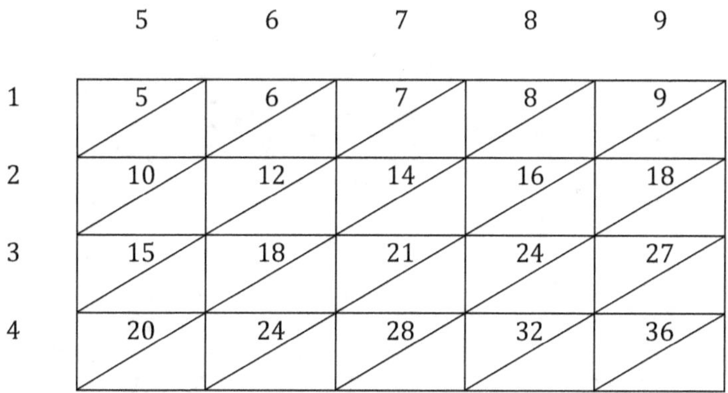

Solution: 1234x56789 = 70077626

EXERCISE PROBLEMS

The following are some of the exercise problems to be solved by using the method of Cell-wise Multiplication to have an acquaintance with the new method of multiplication.

1.	15x11		33.	105x112
2.	11x12		34.	112x121
3.	10x19		35.	106x109
4.	12x14		36.	102x106
5.	45x42		37.	115x112
6.	48x35		38.	120x105
7.	45x48		39.	112x104
8.	37x48		40.	111x111
9.	43x46		41.	112x112
10.	38x48		42.	125x125
11.	38x46		43.	395x59
12.	48x48		44.	498x69
13.	44x44		45.	485x69
14.	39x39		46.	497x79
15.	56x59		47.	593x69
16.	62x56		48.	585x68
17.	57x63		49.	786x89
18.	60x65		50.	888x75
19.	68x64		51.	944x96
20.	57x57		52.	1015x18
21.	55x55		53.	1012x145
22.	64x64		54.	1046x162
23.	95x92		55.	1342x176
24.	98x95		56.	3415x451
25.	85x98		57.	6520x691
26.	97x98		58.	3817x7614
27.	93x96		59.	5711x8711
28.	85x88		60.	6716x9816
29.	86x96		61.	5625x7825
30.	88x88		62.	23451x34567
31.	94x94		63.	45678x76584
32.	99x99			

CHAPTER 4

MODIFIED CELL-WISE MULTIPLICATION

In Chapter 3 we have introduced a simple method of multiplication of two numbers with any number of digits namely, Cell-wise multiplication. This procedure is based on the cell-wise multiplication principles. It does not require any carry over during the multiplication but requires only the multiplication of numbers from 1 to 9. Consequently this procedure is very compact and occupies little space; it is easy to diagnose due to cell-wise multiplication; it is easy to multiply any two numbers with any number of digits. This method is very much useful in situations where the calculators or computers are not available and in some situations where the computers and calculators cannot be used due to large number of digits involved in the numbers to be multiplied. Further it will improve the individual's skill on mathematical computations.

In this chapter some modifications on the method of cell-wise multiplication are introduced and are also illustrated with the help of numerical examples. The modifications on the method of Cell-wise multiplication will improve the speed of the computations. For the sake of convenience of the readers the steps involved in the method of Cell-wise multiplication are presented.

The steps involved in this procedure are as follows:

i.   Let the number of digits in the given numbers be m and n respectively.

ii. Draw mxn array with mn cells.

iii. Write the first number vertically to the left of the array from the top to down such that i[th] digit from left corresponding to the ith row and is denoted by Ri.

iv. Write the second number horizontally to the top of the array from left to right such that the jth digit from left corresponding to the jth column and is denoted by Cj.

v. Find the product of ith digit of first number and j[th] digit of the second number and write the value at the (i,j)[th] cell. That is, the entry at the (i,j)[th] cell is RixCj.

vi. Start from the (m,n)[th] cell, find the sum of the numbers appeared in the secondary diagonals and write the results from right to left.

vii. If any value is more than 10 reduce to mod 10 and carry over the remainder to the next value.

viii. Continue until to reach the cell (1,1).

For more details one may refer the example given below.

Example 4.1: To find the product of 1213x1415, draw an array with 4x4=16 cells as given in the Figure 4.1. Write down the numbers 1213 and 1415 as written in the figure then find the products of 1x1, 1x4, 1x1, 1x5, 2x1, 2x4, 2x1, 2x5 and so on and write the results in their respective cells. Find the sum of the numbers in the secondary diagonals starting from the cell (4,4), one may get 5, 9, 3, 6, 1, 7 and 1. Finally writing these values from right to left one may get the value of 1213x1415 = 1716395 as given in Figure 4.1.

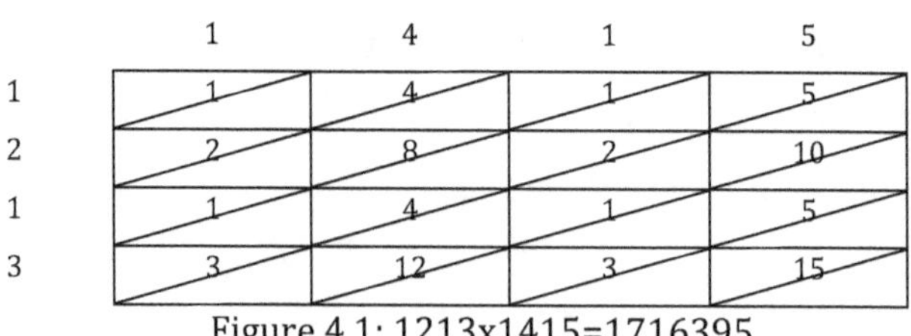

Figure 4.1: 1213x1415=1716395

MODIFIED CELL-WISE MULTIPLICATION

As pointed out in Chapter 3, the method of Cell-wise multiplication is based on the cell-wise multiplication principle; which will have several advantages over the method of conventional multiplication. However the drawback is that it requires mn square cells as an mxn array with m rows and n columns. If m and n becomes large this method requires large number of cells, as a result the speed of computation becomes reduced. To overcome these difficulties, some modifications by the way of combining cells on this method are introduced. Consequently the number of cells is reduced to several folds and the speed of computations is improved considerably.

MODIFICATION I

Suppose that the number of digits involved in the two numbers to be multiplied is a multiple of a certain number then this method is more apt to do the computation with more speed. Let p and q be the number of digits respectively on the two numbers such that $p = mr$ and $q = nr$. The steps involved in the modified Cell-wise multiplication are as follows:

i.   Let the number of digits in the given numbers be $p=mr$ and $q=nr$ respectively.

ii.   Draw mxn array with mn cells.

iii.   Write the first number vertically to the left of the array from the top to down such that from $(r(i-1)+1)^{th}$ digit to ri^{th} digit from left corresponding to the ith row and is denoted by R_i where $i=1,2,...,m$.

iv.   Write the second number horizontally to the top of the array from left to right such that from $(r(j-1)+1)^{th}$ digit to rj^{th} digit from left corresponding to the jth column and is denoted by C_j where $j=1,2,...,n$.

v. Find the product of the corresponding numbers in i^{th} row and in j^{th} column of the array and write the value at the $(i,j)^{th}$ cell. That is, the entry at the $(i,j)^{th}$ cell is $R_i \times C_j$.

vi. Start from the $(m,n)^{th}$ cell, find the sum of the numbers appeared in the secondary diagonals and write the results from right to left.

vii. If any value is more than 10^r reduce to mod 10^r and carry over the remainders to the next value. That is, write r digits at a time.

viii. Continue until to reach the cell $(1,1)$.

For more details one may refer the examples given below.

Example 4.2: Consider the problem given in Example 4.1, to find the product of 1213x1415 which requires drawing 4x4 = 16 cells in a 4x4 array. In this modified method to find the value of 1213x1415, draw an array with 2x2=4 cells as given in the Figure 4.2. Write down the number 1213 as 12 and 13 and the number 1415 as 14 and 15 as written in the figure then find the products of 12x14, 12x15, 13x14 and 13x15 and write the results in their respective cells. Find the sum of the numbers in the secondary diagonals starting from the cell (2,2) and reduce to mod 100 one may get 95,63,71 and 1. Finally writing these values from right to left one may get the value of 1213x1415 =1716395 as given in Figure 4.2, which will agree with the result obtained in Example 4.1.

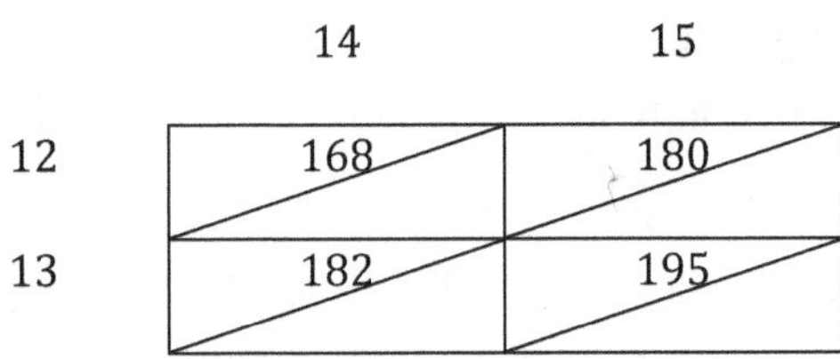

Figure 4.2: 1213x1415= 1716395

Example 4.3: To find the product of 121312x141515 which requires drawing 6x6 = 36 cells in a 6x6 array. In this modified method to find the value of 121312x141515, draw an array with 3x3=9 cells or 2x2=4 cells as given in the Figure 4.3.1 and 4.3.2 respectively. Write down the number 121312 as 12, 13 and 12 and the number 141515 as 14, 15 and 15 as written in the Figure 4.3.1 then find the products of 12x14, 12x15, 12x15, 13x14, 13x15, 13x15, 12x14, 12x15, 12x15 and write the results in their respective cells. Find the sum of the numbers in the secondary diagonals starting from the cell (3, 3) and reduce to mod 100 one may get 80, 76, 46, 67, 71, and 1. Finally writing these values from right to left one may get the value of 121312x141515 =17167467680 as given in Figure 4.3.1.

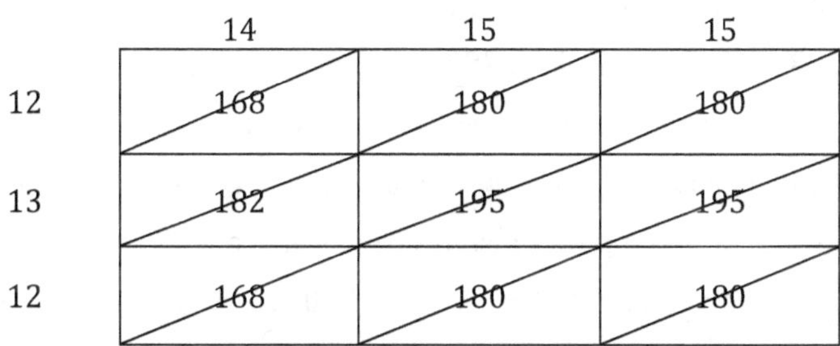

Figure 4.3.1: 121312x141515= 17167467680

Similarly, write down the number 121312 as 121 and 312 and the number 141515 as 141 and 515 as written in the Figure 4.3.2 then find the products of 121x141, 121x515, 312x141 and 312x515. Now write the results in their respective cells. Find the sum of the numbers in the secondary diagonals starting from the cell (2,2) and reduce to mod 1000 one may get 680, 467, 167, and 17. Finally writing these values from right to left one may get the value of 121312x141515 =17167467680 as given in Figure 4.3.2, which will agree with the result obtained in Figure 4.3.1.

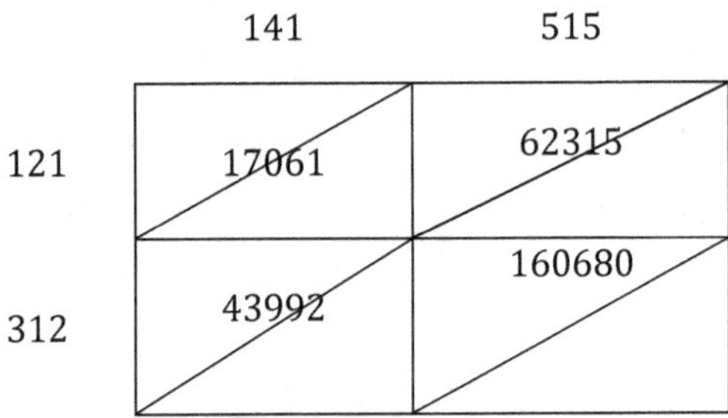

Figure 4.3.2: 121312x141515= 17167467680

Example 4.4: To find the product of 123456789x123456789, draw an array with 3x3=9 cells as given in the Figure 4.4. Write down the number 123456789 as 123, 456 and 789 as written in the Figure 4.4 and then find the products of 123x123, 123x456, 123x789, 456x123, 456x456, 456x789, 789x123, 789x456 and 789x789 and write the results in their respective cells. Find the sum of the numbers in the secondary diagonals starting from the cell (3,3) and reducing mod 1000 one may get 521, 190, 750, 578, 241 and 15. Finally writing these values from right to left one may get the value of 123456789x123456789 =15241578750190521 as given in Figure 4.4, which will agree with the result given in Figure 3.5.

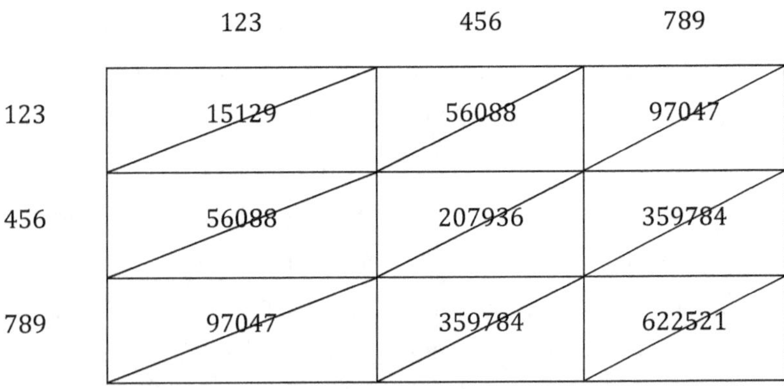

Figure 4.4: 123456789x123456789=15241578750190521

MODIFICATION II

In the modification of cell-wise multiplication discussed above, it is assumed that the number of digits in the two numbers to be multiplied is a multiple of a constant r. That is, the number of digits in the two numbers are p = mr and q = nr respectively. However there are other situations where the number of digits in p and q may not be a multiple of r and hence the modifications suggested in above may not be useful. Consequently, some modifications are suggested to overcome these difficulties and are referred as modifications II. These are illustrated with the help of numerical examples.

Let the number of digits in the two numbers be p=mr+s and q=nr+t respectively. In such situations one may have the following two cases:

Case I: Use of mxn array with mn cells

Case II: Use of (m+1)x(n+1) array with (m+1)(n+1) cells.

In case I, one has to use only an mxn array, allot the first (r+s) digits to the first row and r digits to each of the remaining (m-1) rows as discussed earlier. Similarly allot the first (r+t) digits to the first column and r digits to each of the remaining (n-1) columns as discussed earlier. Once the allotment of the digits to rows and columns is over, proceed to the remaining steps as given in the cell-wise Multiplication procedure.

In case II, one has to use an (m+1)x(n+1) array, allot the first s digits to the first row and r digits to each of the remaining m rows and similarly allot the first t digits to the first column and r digits to each of the remaining n columns as discussed earlier. Once the allotment of the digits to rows and columns is over, proceed to the remaining steps as given in the cell-wise Multiplication procedure.

Example 4.5: To find the product of 12345x1234567 consider the following two cases:

Case I: Draw an array with 2x3=6 cells as given in the Figure 4.5.1. Write down the number 12345 as 123 and 45 and the number 1234567 as 123, 45 and 67 as written in the Figure 4.5.1 and then find the products of 123x123, 123x45, 123x67, 45x123, 45x45, and 45x67 and write the results in their respective cells. Find the sum of the numbers in the secondary diagonals starting from the cell (2,3) and reduce to mod 100 one may get 15, 96, 72, 40, 52 and 1. Finally writing these values from right to left one may get the value of 12345x1234567 =152410729615 as given in Figure 4.5.1.

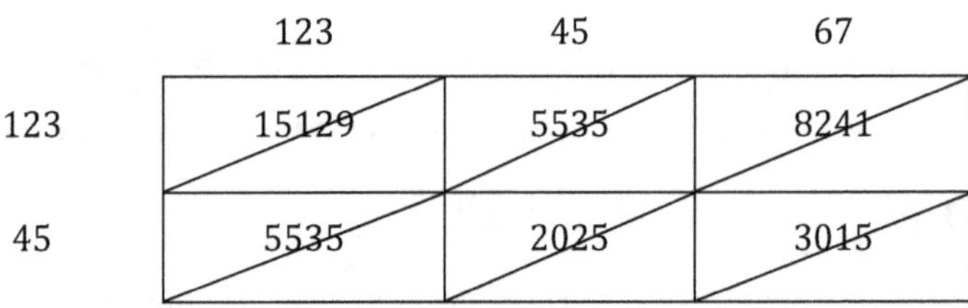

Figure 4.5.1: 12345x1234567=15240729615

Case II: Draw an array with 3x4=12 cells as given in the Figure 4.5.2. Write down the number 12345 as 1, 23 and 45 and the number 1234567 as 1, 23, 45 and 67 as written in the Figure 4.5.2 and then find the products of 1x1, 1x23, 1x45, 1x67, 23x1, 23x23, 23x45, 23x67, 45x1, 45x23, 45x45,and 45x67 and write the results in their respective cells. Find the sum of the numbers in the secondary diagonals starting from the cell (3,4) and reduce to mod 100 one may get 15, 96, 72, 40, 52 and 1. Finally writing these values from right to left one may get the value of 12345x1234567 =152410729615 as given in Figure 4.5.2, which will agree with the result given in Figure 4.5.1.

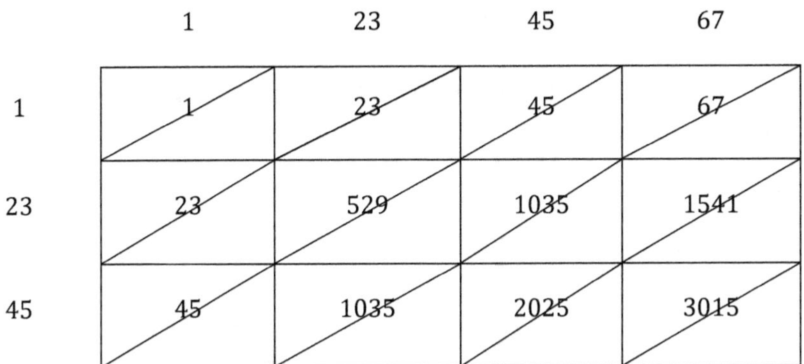

Figure 4.5.2: 12345x1234567=15240729615

SOLVED PROBLEMS

The following are some of the solved problems given to explain the method of Modified Cell-wise Multiplication:

Problem 4.1: To find 125x114

	1	14
1	1	14
25	25	350

Solution: 125x114=14250

Problem 4.2: To find 1622x2412

	24	12
16	384	192
22	528	264

Solution: 1622x2412=3912264

Problem 4.3: To find 512x1814

	18	14
5	90	70
12	216	168

Solution: 512x1814=928768

Problem 4.4: To find 816x214515

	21	45	15
8	168	360	120
16	336	720	240

Solution: 816x214515 =175044240

Problem 4.5: To find 23545x132514

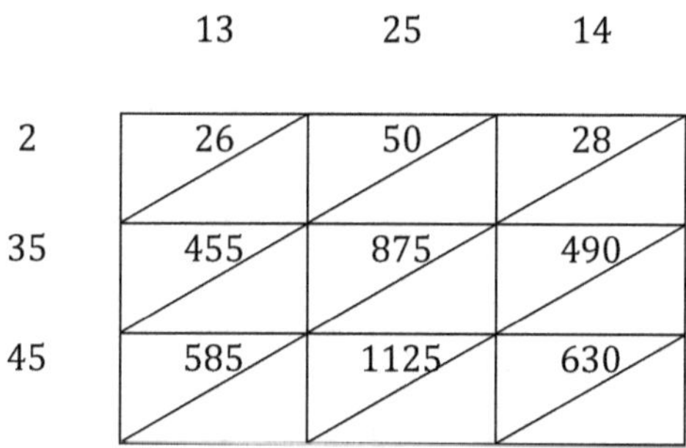

Solution: 23545x132514=3120042130

Problem 4.6: To find 212223x131215

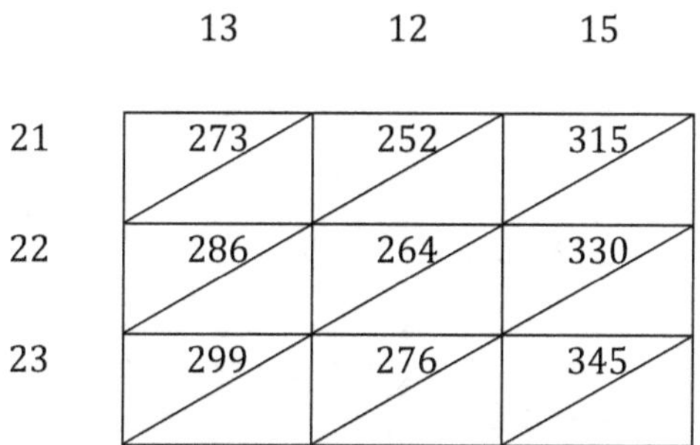

Solution: 212223x131215= 27846840945

Problem 4.7: To find 12125x1125202

Solution: 12125x1125202 = 13643074250

Problem 4.8: To find 125120x51251

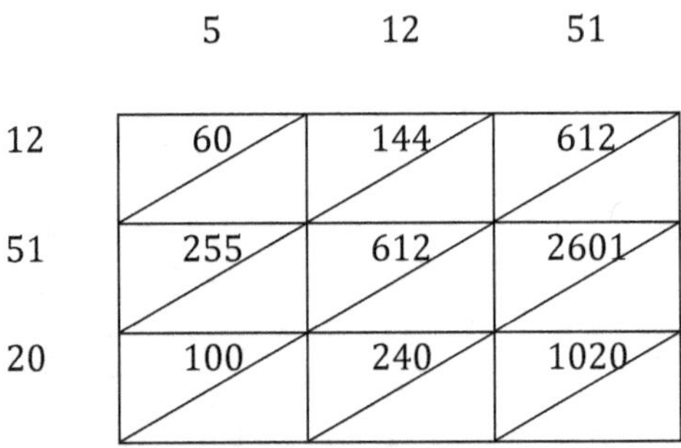

Solution: 125120x51251 = 6412525120

Problem 4.9: To find 51212x15222551

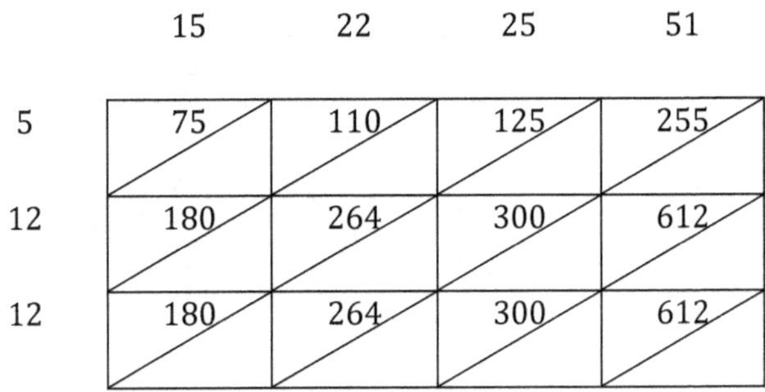

Solution: 51212x15222551=7795772281812

Problem 4.10: To find 12251225x12251225

	1225	1225
1225	1500625	1500625
1225	1500625	1500625

Solution: 12251225x12251225 = 150092514000625

EXERCISE PROBLEMS

The following are some of the exercise problems to be solved by using the method of Modified Cell-wise Multiplication to have an acquaintance with the new method of multiplication.

1. 15678x11234
2. 12345x54321
3. 10101x12059
4. 12243x14567
5. 45987x42678
6. 484848x353535
7. 454545x485648
8. 373737x555555
9. 666666x777777
10. 838383x848484
11. 333333x4444444
12. 444444x8787878
13. 565656x9797979
14. 312312x3999993
15. 656565x5959595
16. 497497497x798798798
17. 593593593x645645645
18. 585585585x688688688
19. 786786786x896896896
20. 888888888x758758758

Chapter 5

Multiplication of Numbers- Based at 10

Chapter 2 through Chapter 4 of this book "Multiplication Made Easy" is dedicated to discuss some simple methods of multiplication of two numbers with any number of digits, which have a lot of advantages over the conventional method of multiplication. In Chapter 2 we have discussed the method of multiplication by line graphs whereas chapters 3 and 4 are dedicated to discuss a more general and elegant procedure of multiplication of two numbers with any number of digits namely, Cell-wise Multiplication and some modifications on the procedure to speed up the computations. These methods are having a lot of advantages over the conventional method of multiplication due to the simplicity and the nature of the principles involved. The prime most importance of these methods is that there is no restriction on the number of digits involved in the numbers to be multiplied. For example if the number of digits involved in the numbers is more than 6 or 7 digits it is very difficult to use the calculators and the computers to get the accurate value of the product of such numbers where as the methods discussed in Chapter 3 and 4 can be easily adopted even in such situations.

In this Chapter it is planned to discuss the multiplication of two numbers based at 10 and the procedure is illustrated with the help of several numerical examples. Even though the method is very simple in nature it will certainly help the youngsters to understand what are we actually doing during the multiplication? As mentioned in the introduction several simplified procedures for multiplication of two numbers are to be discussed, which are based at 10, 50, 100, 500, 1000, etc. For example, the methods based at n means the numbers to be multiplied are very close to either sides of the value n. If the two numbers to be multiplied are very close to some numbers such as 10, 50, 100, 500, 1000, etc. one can easily obtain the value of the multiplication mentally. In this procedure one may come across the following two cases:

i.   Both the numbers to be multiplied are less (greater) than the number n and

ii. One of the two numbers is greater than n and the other one is less than n.

The other details of working procedures of these methods are explained in appropriate headings and also illustrated with the help of several numerical examples.

MULTIPLICATION BASED AT 10

The procedure of multiplication based at 10 is explained for finding the multiplication of two numbers, which are very close to either side of the value 10 and can be easily extended for other numbers like 100, 1000, 10000 etc., As mentioned above one may have the following two cases. The steps involved are as follows:

Case I: When the numbers are less (greater) than 10.

i.   Let X_1 and X_2 be the two numbers to be multiplied. Write down the numbers vertically one by one as written in Example 5.1.

ii. Subtract 10 from the numbers and write the remainders say R_1 and R_2 at the right hand side.

iii. Multiply R_1 and R_2 and write down the results as done in the Example 5.1.

iv. Add R_2 with X_1 or R_1 with X_2 and write down the results as done in Example 5.1.

v.   If any of the sums is more than 10 then reduce to mod 10 and carry over the remainders to the next position.

vi. Write the digits one at a time from right to left to get the result.

For more details please refer the numerical examples given below:

Example 5.1: To find the product of 12x13, write down the numbers 12 and 13 vertically as written in the Figure 5.1. Then subtract the value 10 from the numbers and write the remainders 2 and 3 at the right hand side corresponding to the respective values 12 and 13. Find the product of 2 and 3 and write down the resulting number 6 vertically below to the numbers 2 and 3. Now add the value 3 to 12 or 2 to 13; both will yield the same results and write the resulting number 15 vertically below to the numbers 12 and 13. Now write these values from right to left one digit at a time. Carry over the remainders after reducing to mod 10 if any of the values is more than 10, which yield 6, 5 and 1 and write these numbers from right to left to get the product of 12x13=156.

$$
\begin{array}{c|c}
12 & +2 \\
13 & +3 \\
\hline
15 & 6
\end{array}
$$

$$(15)(6)=>156$$

Figure 5.1:12x13 = 156

Example 5.2: To find the product of 8x9, write down the numbers 8 and 9 vertically as written in the Figure 5.2. Then subtract the value 10 from the two numbers 8 and 9 and write the remainders -2 and -1 at the right hand side corresponding to the respective values 8 and 9. Find the product of -2 and -1 and write down the resulting number 2 vertically below numbers -2 and -1. Now add the value -1 to 8 or -2 to 9; both will yield the same result and write the resulting number 7 vertically below to the numbers 8 and 9. Now write these values from right to left one digit at a time. Carry over the remainders after reducing to mod 10 if any value is more than 10, which yield 2 and 7 and write these numbers from right to left to get the product of 8x9=72.

$$
\begin{array}{c|c}
8 & -2 \\
9 & -1 \\
\hline
7 & 2
\end{array}
$$

$$(7)(2)=>72$$

Figure 5.2: 8x9 = 72

Example 5.3: To find the product of 21x13, write down the numbers 21 and 13 vertically as written in the Figure 5.3. Then subtract the value 10 from 21 and 13 and write the remainders 11 and 3 at the right hand side corresponding to the respective values 21 and 13. Find the product of 11 and 3 and write down the resulting number 33 vertically below to the numbers 11 and 3. Now add the value 3 to 21 or 11 to 13; both will yield the same result and write the resulting number 24 vertically below to the numbers 21 and 13. Now write these values from right to left one digit at a time. Carry over the remainders after reducing to mod 10 if any value is more than 10, which yield 3, 7 and 2 and write these numbers from right to left to get the product of 21x13=273

$$
\begin{array}{c|c}
21 & +11 \\
13 & +3 \\
\hline
24 & 33
\end{array}
$$

$$(24)(33)=>(24+3)3=(27)(3)=273$$

Figure 5.3: 21x13 = 273

Case II: When one of the two numbers is greater than 10 and other one is less than 10

i.   Let X_1 and X_2 be the two numbers to be multiplied. Write down the numbers vertically one by one as written in Example 5.4.

ii.   Subtract 10 from the numbers and write the remainders say R_1 and R_2 at the right hand side.

iii.   Multiply R_1 and R_2 and write down the results as done in the Example 5.4.

iv.   Add R_2 with X_1 or R_1 with X_2 and write down the results as done in Example 5.4

v.   Since the product of R_1 and R_2 is negative take the complement of the last digit and if the absolute value is more than 10 then reduce to mod 10 and carry over the negative value of remainders to the next position.

vi.   Write the digits one at a time from right to left to get the required result.

For more details please refer the numerical examples given below:

Example 5.4: To find the product of 8x13, write down the numbers 8 and 13 vertically as written in the Figure 5.4. Then subtract the value 10 from 8 and 13 and write the remainders -2 and 3 at the right hand side corresponding to the respective values 8 and 13. Find the product of -2 and 3 and write down the resulting number -6 vertically below to the numbers -2 and 3. Now add the value 3 to 8 or -2 to 13; both will yield the same result and write the resulting number 11 vertically below to the numbers 8 and 13. Now as stated in step v, write these values from right to left one digit at a time and carry over the remainders after reducing to mod 10 if any value is more than 10, which yield 4, 0 and 1 and write these numbers from right to left to get the product of 8x13=104

$$
\begin{array}{r|r}
8 & -2 \\
13 & +3 \\
\hline
11 & -6 \\
\end{array}
$$

(11)(-6)=>(11-1)(4)=(10)(4)=104

Figure 5.4: 8x13 = 104

Example 5.5: To find the product of 8x16, write down the numbers 8 and 16 vertically as written in the Figure 5.5. Then subtract the value 10 from the numbers 8 and 16 and write the remainders -2 and 6 at the right hand side corresponding to the respective values 8 and 16. Find the product of -2 and 6 and write down the resulting number -12 vertically below to the numbers -2 and 6. Now add the value 6 to 8 or -2 to 16; both will yield the same result and write the resulting number 14 vertically below to the numbers 8 and 16. Now as stated in step v, write these values from right to left one digit at a time. Carry over the remainders after reducing to mod 10 if any value is more than 10, which yield 8, 2 and 1. Write down these numbers from right to left to get the product of 8x16=128

$$
\begin{array}{r|r}
8 & -2 \\
16 & +6 \\
\hline
14 & -12 \\
\end{array}
$$

(14)(-12)=>(13)(-2)=(12)(8) = 128

Figure 5.5: 8x16 = 128

SOLVED PROBLEMS

The following are some of the solved problems given to explain the method of Multiplication Based at 10.

Problem 5.1: To find 15x14

15	+5
14	+4
19	20

$$(19)(20) => (19+2)(0) = (21)(0) = 210$$

Solution: 15x14 = 210

Problem 5.2: To find 17x13

17	+7
13	+3
20	21

$$(20)(21) => (20+2)(1) = (22)(1) = 221$$

Solution: 17x13 = 221

Problem 5.3: To find 15x17

15	+5
17	+7
22	35

$$(22)(35) => (22+3)(5) = (25)(5) = 255$$

Solution: 15x17 = 255

Problem 5.4: To find 15x9

$$
\begin{array}{c|c}
15 & +5 \\
9 & -1 \\
\hline
14 & -5
\end{array}
$$

$$(14)(-5)=>(14-1)(5)=(13)(5) = 135$$

Solution: 15x9 = 135

Problem 5.5: To find 17x8

$$
\begin{array}{c|c}
17 & +7 \\
8 & -2 \\
\hline
15 & -14
\end{array}
$$

$$(15)(-14)=>(15-1)(-4)=(13)(6) = 136$$

Solution: 17x8 = 136

Problem 5.6: To find 19x9

$$
\begin{array}{c|c}
19 & +9 \\
9 & -1 \\
\hline
18 & -9
\end{array}
$$

$$(18)(-9)=>(18-1)(1)=(17)(1) = 171$$

Solution: 19x9 = 171

Exercise Problems

The following are some of the exercise problems to be solved by using the method of Multiplication Based at 10 to have an acquaintance with the new method of multiplication.

1.	15x11		16.	5x18
2.	11x12		17.	6x19
3.	10x19		18.	8x18
4.	12x16		19.	9x14
5.	15x17		20.	9x16
6.	12x15		21.	9x23
7.	12x14		22.	7x13
8.	11x11		23.	9x26
9.	17x17		24.	8x24
10.	15x15		25.	8x28
11.	9x8		26.	7x16
12.	8x5		27.	21x23
13.	8x8		28.	24x28
14.	9x9		29.	25x21
15.	7x9		30.	15x21

CHAPTER 6

MULTIPLICATION OF NUMBERS- BASED AT 100

In Chapter 5 we have discussed the multiplication of two numbers based at 10 and the procedure is also illustrated with the help of several numerical examples. That is, the two numbers to be multiplied are very close to 10 in either side the above method is used to find their products. Initially one has to do the steps given in the procedure and once get familiar with this method no need to do all the steps and one can do the multiplications mentally. The procedure discussed for the multiplication based at 10 can be easily extended to do multiplication of two numbers, which are very close to the value 100. In this procedure one may come across the following two cases:

i.   Both the numbers to be multiplied are less (greater) than the number 100 and

ii. One of the two numbers is greater than n and the other one is less than 100.

The other details of working procedures of these methods are explained in appropriate headings and also illustrated with the help of several numerical examples.

MULTIPLICATION BASED AT 100

The procedure of multiplication based at 100 is explained for finding the multiplication of two numbers, which are very close to either side of the value 100. As mentioned above one may have the following two cases.

The steps involved are as follows:

Case I: When the numbers are less (greater) than 100.

i. Let X_1 and X_2 be the two numbers to be multiplied. Write down the numbers vertically one by one as written in Example 6.1.

ii. Subtract 100 from the numbers and write the remainders say R_1 and R_2 at the right hand side.

iii. Multiply R_1 and R_2. If any single digit comes, convert it to a two digits number by adding zero at the prefix. Write down the results as done in the Example 6.1.

iv. Add R_2 with X_1 or R_1 with X_2 and write down the results as done in Example 6.1.

v. If any of the sums is more than 100 then reduce to mod 100 and carry over the remainders to the next position.

vi. Write two digits at a time from right to left to get the result.

For more details please refer the numerical examples given below:

Example 6.1: To find the product of 102x113, write down the numbers 102 and 113 vertically as written in the Figure 6.1. Then subtract the value 100 from the numbers and write the remainders 2 and 13 at the right hand side corresponding to the respective values 102 and 113. Find the product of 2 and 13 and write down the resulting number 26 vertically below to the numbers 2 and 13. Now add the value 13 to 102 or 2 to 113; both will yield the same results and write the resulting number 115 vertically below to the numbers 102 and 113. Now write these values from right to left two digits at a time. Carry over the remainders after reducing to mod 100 if any of the values is more than 100, which yield 26, 15 and 1 and write these numbers from right to left to get the product of 102x113=11526.

$$
\begin{array}{c|c}
102 & +\,2 \\
113 & +13 \\
\hline
115 & 26
\end{array}
$$

$$(115)(26) => 11526$$

Figure 6.1:102x113 = 11526

Example 6.2: To find the product of 98x89, write down the numbers 98 and 89 vertically as written in the Figure 6.2. Then subtract the value 100 from the two numbers 98 and 89 and write the remainders -2 and -11 at the right hand side corresponding to the respective values 98 and 89. Find the product of -2 and -11 and write down the resulting number 22 vertically below numbers -2 and -11. Now add the value -11 to 98 or -2 to 89; both will yield the same result and write the resulting number 87 vertically below to the numbers 98 and 89. Now write these values from right to left two digits at a time. Carry over the remainders after reducing to mod 100 if any value is more than 100, which yield 22 and 87 and write these numbers from right to left to get the product of 98x89=8722.

$$
\begin{array}{c|c}
98 & -2 \\
89 & -11 \\
\hline
87 & 22
\end{array}
$$

$$(87)(22)=8722$$

Figure 6.2: 98x89 = 8722

Example 6.3: To find the product of 121x113, write down the numbers 121 and 113 vertically as written in the Figure 6.3. Then subtract the value 100 from 121 and 113 and write the remainders 21 and 13 at the right hand side corresponding to the respective values 121 and 113. Find the product of 21 and 13 and write down the resulting number 273 vertically below to the numbers 21 and 13. Now add the value 13 to 121 or 21 to 113; both will yield the same result and write the resulting number 134 vertically below to the numbers 121 and 113. Now write these values from right to left two digits at a time. Carry over the remainders after reducing to mod 100 if any value is more than 100, which yield 73, 36 and 1 and write these numbers from right to left to get the product of 121x113=13673

$$
\begin{array}{c|c}
121 & +21 \\
113 & +13 \\
\hline
134 & 273
\end{array}
$$

$$(134)(273)=>(134+2)(73)=(136)(73)=13673$$

Figure 6.3: 121x113 = 13673

Example 6.4: To find the product of 102x103, write down the numbers 102 and 103 vertically as written in the Figure 6.4. Then subtract the value 100 from the numbers and write the remainders 2 and 3 at the right hand side corresponding to the respective values 102 and 103. Find the product of 2 and 3 and write down the resulting number 6 vertically below to the numbers 2 and 3. Now add the value 3 to 102 or 2 to 103; both will yield the same results and write the resulting number 105 vertically below to the numbers 102 and 103. Now write these values from right to left two digits at a time. Carry over the remainders after reducing to mod 100 if any of the values is more than 100, which yield 06, 05 and 1 and write these numbers from right to left to get the product of 102x103=10506

$$
\begin{array}{c|c}
102 & +2 \\
103 & +3 \\
\hline
105 & 6
\end{array}
$$

$$(105)(6)=>(105)(06)=10506$$

Figure 6.4:102x103 = 10506

Case II: When one of the two numbers is greater than 100 and the other one is less than 100

i.   Let X_1 and X_2 be the two numbers to be multiplied. Write down the numbers vertically one by one as written in Example 6.5.

ii. Subtract 100 from the numbers and write the remainders say R_1 and R_2 at the right hand side.

iii. Multiply R_1 and R_2. If any single digit comes, convert it to a two digits number by adding zero at the prefix. Write down the results as done in the Example 6.5.

iv. Add R_2 with X_1 or R_1 with X_2 and write down the results as done in Example 6.5.

v.   Since the product of R_1 and R_2 is negative take the complement of the last two digits and if the absolute value is more than 100 then reduce to mod 100 and carry over the negative value of the remainders to the next position.

vi. Write two digits at a time from right to left to get the required result.

For more details please refer the numerical examples given below:

Example 6.5: To find the product of 98x113, write down the numbers 98 and 113 vertically as written in the Figure 6.5. Then subtract the value 100 from 98 and 113 and write the remainders -2 and 13 at the right hand side corresponding to the respective values 98 and 113. Find the product of -2 and 13 and write down the resulting number -26 vertically below to the numbers -2 and 13. Now add the value 13 to 98 or -2 to 113; both will yield the same result and write the resulting number 111 vertically below to the numbers 98 and 113. Now as stated in step v, write these values from right to left two digits at a time. Carry over the remainders after reducing to mod

100 if any value is more than 100, which yield 74, 10 and 1 and write these numbers from right to left to get the product of 98x113=11074

$$\begin{array}{c|c} 98 & -2 \\ 113 & +13 \\ \hline 111 & -26 \end{array}$$

(11)(-26)=>(111-1)(74)=(110)(74)=11074

Figure 6.5: 98x113 = 11074

Example 6.6: To find the product of 88x106, write down the numbers 88 and 106 vertically as written in the Figure 6.6. Then subtract the value 100 from the numbers 88 and 106 and write the remainders -12 and 6 at the right hand side corresponding to the respective values 88 and 106. Find the product of -12 and 6 and write down the result -72 vertically below to the numbers -12 and 6. Now add the value 6 to 88 or -12 to 106; both will yield the same result and write the resulting number 94 vertically below to the numbers 88 and 106. Now as stated in step v, write these values from right to left two digits at a time. Carry over the remainders after reducing to mod 100 if any value is more than 100, which yield 28 and 93. Write down these numbers from right to left to get the product of 88x106=9328

$$\begin{array}{c|c} 88 & -12 \\ 106 & +6 \\ \hline 94 & -72 \end{array}$$

(94)(-72)=>(94-1)(28)=(93)(28) = 9328

Figure 6.6: 88x106 = 9328

SOLVED PROBLEMS

The following are some of the solved problems given to explain the method of Multiplication based at 100.

Problem 6.1: To find 105x104

$$
\begin{array}{c|c}
105 & +5 \\
104 & +4 \\
\hline
109 & 20
\end{array}
$$

$$(109)(20)=10920$$

Solution: 105x104 = 10920

Problem 6.2: To find 107x103

$$
\begin{array}{c|c}
107 & +7 \\
103 & +3 \\
\hline
110 & 21
\end{array}
$$

$$(110)(21)= 11021$$

Solution: 107x103 = 11021

Problem 6.3: To find 115x117

$$
\begin{array}{c|c}
115 & +15 \\
117 & +17 \\
\hline
132 & 255
\end{array}
$$

$$(132)(255)=>(132+2)(55)=(134)(55) = 13455$$

Solution: 115x117 = 13455

Problem 6.4: To find 115x99

$$\begin{array}{c|c} 115 & +15 \\ 99 & -1 \\ \hline 114 & -15 \end{array}$$

$$(114)(-15)=>(114-1)(85)=(113)(85) = 11385$$

Solution: 115x99 = 11385

Problem 6.5: To find 117x88

$$\begin{array}{c|c} 117 & +17 \\ 88 & -12 \\ \hline 105 & -204 \end{array}$$

$$(105)(-204)=>(105-2)(-04)=(103)(-04)=(102)(96)=10296$$

Solution: 117x88 = 10296

Problem 6.6: To find 109x89

$$\begin{array}{c|c} 109 & +9 \\ 89 & -11 \\ \hline 98 & -99 \end{array}$$

$$(98)(-99)=>(98-1)(01)=(97)(01) = 9701$$

Solution: 109x89 = 9701

EXERCISE PROBLEMS

The following are some of the exercise problems to be solved by using the method of Multiplication Based at 100 to have an acquaintance with the new method of multiplication.

1.	115x111		16.	95x118
2.	111x112		17.	116x99
3.	110x109		18.	98x118
4.	112x106		19.	89x114
5.	115x117		20.	97x116
6.	112x115		21.	95x112
7.	121x114		22.	87x113
8.	116x119		23.	92x126
9.	107x117		24.	88x114
10.	115x115		25.	89x118
11.	97x89		26.	97x119
12.	88x95		27.	121x123
13.	87x85		28.	124x118
14.	95x92		29.	125x121
15.	97x90		30.	115x127

CHAPTER 7

MULTIPLICATION OF NUMBERS- BASED AT 1000

In Chapters 5 and 6 we have discussed the multiplication of two numbers based at 10 and 100 respectively and are also illustrated with the help of several numerical examples. That is, the two numbers to be multiplied are very close to 10 or 100 in either side of the numbers then the above method is used to find their products. The procedure discussed for the multiplication based at 10 and 100 can be easily extended to do multiplication of two numbers, which are very close to the value 1000. In the procedure of multiplication based at 10, one has to write one digit at a time whereas in the multiplication based at 100 one has to write three digits at a time. Similarly in the case of multiplication based at 1000 one has to write 3 digits at a time. Initially one has to do the steps given in the procedure and once get familiar with this method no need to do all the steps and one can do the multiplications mentally. In this procedure one may come across the following two cases:

i.   Both the numbers to be multiplied are less (greater) than the number 1000 and

ii. One of the two numbers is greater than n and the other one is less than 1000.

The other details of working procedures of these methods are explained in appropriate headings and also illustrated with the help of several numerical examples.

MULTIPLICATION BASED AT 1000

The procedure of multiplication based at 1000 is explained for finding the multiplication of two numbers, which are very close to either side of the value 1000. As mentioned above one may have the following two cases. The steps involved are as follows:

Case I: When the numbers are less (greater) than 1000.

i. Let X_1 and X_2 be the two numbers to be multiplied. Write down the numbers vertically one by one as written in Example 7.1

ii. Subtract 1000 from the numbers and write the remainders say R_1 and R_2 at the right hand side.

iii. Multiply R_1 and R_2. If the value of $R_1 x R_2$ has less than three digits, convert it to a three digits number by adding required number of zeros at the prefix. Write down the results as done in the Example 7.1.

iv. Add R_2 with X_1 or R_1 with X_2 and write down the result as done in Example 7.1.

v. If any of the sums is more than 1000 then reduce to mod 1000 and carry over the remainders to the next position.

vi. Write three digits at a time from right to left to get the results.

For more details please refer the numerical examples given below:

Example 7.1: To find the product of 1012x1013, write down the numbers 1012 and 1013 vertically as written in the Figure 7.1. Then subtract the value 1000 from the numbers and write the remainders 12 and 13 at the right hand side corresponding to the respective values 1012 and 1013. Find the product of 12 and 13 and write down the resulting number 156 vertically below to the numbers 12 and 13. Now add the value 13 to 1012 or

12 to 1013; both will yield the same results and write the resulting number 1025 vertically below to the numbers 1012 and 1013. Now write these values from right to left three digits at a time. Carry over the remainders after reducing to mod 1000 if any of the values is more than 1000, which yield 156, 025 and 1 and write these numbers from right to left to get the product of 1012x1013=1025156

$$
\begin{array}{c|c}
1012 & +12 \\
1013 & +13 \\
\hline
1025 & 156 \\
\end{array}
$$

$$(1025)(156)=>1025156$$

Figure 7.1:1012x1013 = 1025156

Example 7.2: To find the product of 998x989, write down the numbers 998 and 989 vertically as written in the Figure 7.2. Then subtract the value 1000 from the two numbers 998 and 989 and write the remainders -2 and -11 at the right hand side corresponding to the respective values 998 and 989. Find the product of -2 and -11 and write down the resulting number 22 vertically below numbers -2 and -11. Now add the value -11 to 998 or -2 to 989; both will yield the same result and write the resulting number 987 vertically below to the numbers 998 and 989. Now write these values from right to left three digits at a time. Carry over the remainders after reducing to mod 1000 if any value is more than 1000, which yield 022 and 987 and write these numbers from right to left to get the product of 998x989=987022.

$$
\begin{array}{c|c}
998 & -2 \\
989 & -11 \\
\hline
987 & 22 \\
\end{array}
$$

$$(987)(22)=(987)(022)=987022$$

Figure 7.2: 998x989 = 987022

Example 7.3: To find the product of 1021x1013, write down the numbers 1021 and 1013 vertically as written in the Figure 7.3. Then subtract the value 1000 from 1021 and 1013 and write the remainders 21 and 13 at the right hand side corresponding to the respective values 1021 and 1013. Find the product of 21 and 13 and write down the resulting number 273 vertically below to the numbers 21 and 13. Now add the value 13 to 1021 or 21 to 1013; both will yield the same result and write the resulting number 1034 vertically below to the numbers 1021 and 1013. Now write these values from right to left three digits at a time. Carry over the remainders after reducing to mod 1000 if any value is more than 1000, which yield 273, 034 and 1 and write these numbers from right to left to get the product of 1021x1013=1034273.

$$\begin{array}{r|l} 1021 & +21 \\ 1013 & +13 \\ \hline 1034 & 273 \end{array}$$

$$(1034)(273)=1034273$$

Figure 7.3: 1021x1013 = 1034273

Example 7.4: To find the product of 1002x1003, write down the numbers 1002 and 1003 vertically as written in the Figure 7.4. Then subtract the value 1000 from the numbers and write the remainders 2 and 3 at the right hand side corresponding to the respective values 1002 and 1003. Find the product of 2 and 3 and write down the resulting number 6 vertically below to the numbers 2 and 3. Now add the value 3 to 1002 or 2 to 1003; both will yield the same results and write the resulting number 1005 vertically below to the numbers 1002 and 1003. Now write these values from right to left three digits at a time. Carry over the remainders after reducing to mod 1000 if any of the values is more than 1000, which yield 006, 005 and 1 and write these numbers from right to left to get the product of 1002x1003=1005006

$$\begin{array}{c|c}
1002 & +2 \\
1003 & +3 \\
\hline
1005 & 6
\end{array}$$

$$(1005)(6) => (1005)(006) = 1005006$$

Figure 7.4:1002x1003 = 1005006

Case II: When one of the two numbers is greater than 1000 and the other one is less than 1000

i.   Let X_1 and X_2 be the two numbers to be multiplied. Write down the numbers vertically one by one as written in Example 7.5.

ii.   Subtract 1000 from the numbers and write the remainders say R_1 and R_2 at the right hand side.

iii.   Multiply R_1 and R_2. If the value of $R_1 x R_2$ has less than three digits, convert it to a three digits number by adding required number of zeros at the prefix. Write down the results as done in the Example 7.5.

iv.   Add R_2 with X_1 or R_1 with X_2 and write down the results as done in Example 7.5.

v.   Since the product of R_1 and R_2 is negative take the complement of the last three digits and if the absolute value is more than 1000 then reduce to mod 1000 and carry over the negative value of the remainders to the next position.

vi.   Write three digits at a time from right to left to get the required result.

For more details please refer the numerical examples given below:

Example 7.5: To find the product of 998x1013, write down the numbers 998 and 1013 vertically as written in the Figure 7.5. Then subtract the value 1000 from 998 and 1013 and write the remainders -2 and 13 at the right hand side corresponding to the respective values 998 and 1013. Find the product of -2 and 13 and write down the resulting number -26 vertically below to the numbers -2 and 13. Now add the value 13 to 998 or -2 to 1013; both will yield the same result and write the resulting number 1011 vertically below to the numbers 998 and 1013. Now as stated in step v, write these values from right to left three digits at a time. Carry over the remainders after reducing to mod 1000 if any value is more than 1000, which yield 974, 010 and 1 and write these numbers from right to left to get the product of 998x1013=1010974

$$
\begin{array}{c|c}
998 & -2 \\
1013 & +13 \\
\hline
1011 & -26
\end{array}
$$

$$(1011)(-26) => (1011)(-026) = (1010)(974) = 1010974$$

Figure 7.5: 998x1013 = 1010974

Example 7.6: To find the product of 988x1006, write down the numbers 988 and 1006 vertically as written in the Figure 7.6. Then subtract the value 1000 from the numbers 988 and 1006 and write the remainders -12 and 6 at the right hand side corresponding to the respective values 988 and 1006. Find the product of -12 and 6 and write down the result -72 vertically below to the numbers -12 and 6. Now add the value 6 to 988 or -12 to 1006; both will yield the same result and write the resulting number 994 vertically below to the numbers 988 and 1006. Now as stated in step v, write these values from right to left three digits at a time. Carry over the remainders after reducing to mod 1000 if any value is more than 1000, which yield 928 and 993. Write down these numbers from right to left to get the product of 988x1006=9328

$$
\begin{array}{c|c}
988 & -12 \\
1006 & +6 \\
\hline
994 & -72
\end{array}
$$

$$(994)(-72) => (994)(-072) = (993)(928) = 993928$$

Figure 7.6: 988x1006 = 993928

SOLVED PROBLEMS

The following are some of the solved problems given to explain the method of Multiplication based at 1000.

Problem 7.1: To find 1005x1004

1005	+5
1004	+4
1009	20

$(1009)(20)=>(1009)(020)=1009020$

Solution: 1005x1004 = 1009020

Problem 7.2: To find 1007x1003

1007	+7
1003	+3
1010	21

$(1010)(21)=> (1010)(021)= 1010021$

Solution: 1007x1003 = 1010021

Problem 7.3: To find 1015x1017

1015	+15
1017	+17
1032	255

$(1032)(255)=>(1032)(255)= 1032255$

Solution: 1015x1017 = 1032255

Problem 7.4: To find 1015x999

1015	+15
999	-1
1014	-15

$(1014)(-15)=>(1014)(-015)=(1013)(985)=1013985$

Solution: 1015x999 = 1013985

Problem 7.5: To find 1017x988

1017	+17
988	-12
1005	-204

$(1005)(-204)=>(1004)(796)=1004796$

Solution: 1017x988 = 1004796

Problem 7.6: To find 1009x989

1009	+9
989	-11
998	-99

$(998)(-99)=>(998)(-099)=(997)(901) = 997901$

Solution: 1009x989 = 997901

EXERCISE PROBLEMS

The following are some of the exercise problems to be solved by using the method of Multiplication Based at 1000 to have an acquaintance with the new method of multiplication.

1.	1015x1011		16.	995x1018
2.	1011x1012		17.	1016x996
3.	1010x1009		18.	998x1018
4.	1012x1016		19.	989x1014
5.	1015x1017		20.	997x1016
6.	1012x1015		21.	995x1012
7.	1021x1014		22.	987x1013
8.	1016x1019		23.	992x1026
9.	1007x1017		24.	988x1014
10.	1015x1015		25.	1009x1018
11.	997x989		26.	1007x1019
12.	988x995		27.	1021x1023
13.	987x985		28.	1024x1018
14.	995x992		29.	1025x1021
15.	997x990		30.	1015x1027

CHAPTER 8

MULTIPLICATION OF NUMBERS- BASED AT 10000

In the preceding Chapters we have discussed the multiplication of two numbers based at 10, 100 and 1000 and are also illustrated with the help of several numerical examples. That is, the two numbers to be multiplied are very close to 10, 100 and 1000 in either side of the numbers then the above method is used to find their products. The procedure discussed for the multiplication based at 10, 100, 1000, etc. can be easily extended to do multiplication of two numbers, which are very close to the value 10^k, where k is any positive integer. In the procedure of multiplication based at 10, one has to write one digit at a time; in the multiplication based at 100 one has to write two digits at a time whereas in the case of multiplication based at 10^k one has to write k-digits at a time. That is, by definition of this procedure in the case of multiplication based at 10000 one has to write four digits at a time. In this procedure one may come across the following two cases:

i.   Both the numbers to be multiplied are less (greater) than the number 10000 and

ii.   One of the two numbers is greater than n and the other one is less than 10000.

The other details of working procedures of these methods are explained in appropriate headings and also illustrated with the help of several numerical examples.

MULTIPLICATION BASED AT 10000

The procedure of multiplication based at 10000 is explained for finding the multiplication of two numbers, which are very close to either side of the value 10000. As mentioned above one may have the following two cases. The steps involved are as follows:

Case I: When the numbers are less (greater) than 10000.

i. Let X_1 and X_2 be the two numbers to be multiplied. Write down the numbers vertically one by one as written in Example 8.1

ii. Subtract 10000 from the numbers and write the remainders say R_1 and R_2 at the right hand side.

iii. Multiply R_1 and R_2. If the value of $R_1 x R_2$ has less than four digits, convert it to a four digits number by adding required number of zeros at the prefix. Write down the results as done in the Example 8.1.

iv. Add R_2 with X_1 or R_1 with X_2 and write down the results as done in Example 8.1.

v. If any of the sums is more than 10000 then reduce to mod 10000 and carry over the remainders to the next position.

vi. Write four digits at a time from right to left to get the results.

For more details please refer the numerical examples given below:

Example 8.1: To find the product of 10012x10013, write down the numbers 10012 and 10013 vertically as written in the Figure 8.1. Then subtract the value 10000 from the numbers and write the remainders 12 and 13 at the right hand side corresponding to the respective values 10012 and 10013. Find the product of 12 and 13 and write down the resulting number 156 vertically below to the numbers 12 and 13. Now add the value

13 to 10012 or 12 to 10013; both will yield the same results and write the resulting number 10025 vertically below to the numbers 10012 and 10013. Now write these values from right to left four digits at a time. Carry over the remainders after reducing to mod 10000 if any of the values is more than 10000, which yield 0156, 0025 and 1 and write these numbers from right to left to get the product of 10012x10013=100250156

$$
\begin{array}{c|c}
10012 & +12 \\
10013 & +13 \\
\hline
10025 & 156 \\
\end{array}
$$

(10025)(156)=> (10025)(0156)=100250156

Figure 8.1:10012x10013 = 100250156

Example 8.2: To find the product of 9998x9989, write down the numbers 9998 and 9989 vertically as written in the Figure 8.2. Then subtract the value 10000 from the two numbers 9998 and 9989 and write the remainders -2 and -11 at the right hand side corresponding to the respective values 9998 and 9989. Find the product of -2 and -11 and write down the resulting number 22 vertically below numbers -2 and -11. Now add the value -11 to 9998 or -2 to 9989; both will yield the same result and write the resulting number 9987 vertically below to the numbers 9998 and 9989. Now write these values from right to left four digits at a time. Carry over the remainders after reducing to mod 10000 if any value is more than 10000, which yield 0022 and 0987 and write these numbers from right to left to get the product of 9998x9989=99870022.

$$
\begin{array}{c|c}
9998 & -2 \\
9989 & -11 \\
\hline
9987 & 22 \\
\end{array}
$$

(9987)(22)=>(9987)(0022)=99870022

Figure 8.2: 9998x9989 = 99870022

Example 8.3: To find the product of 10021x10013, write down the numbers 10021 and 10013 vertically as written in the Figure 8.3. Then subtract the value 10000 from 10021 and 10013 and write the remainders 21 and 13 at the right hand side corresponding to the respective values 10021 and 10013. Find the product of 21 and 13 and write down the resulting number 273 vertically below to the numbers 21 and 13. Now add the value 13 to 10021 or 21 to 10013; both will yield the same result and write the resulting number 10034 vertically below to the numbers 10021 and 10013. Now write these values from right to left four digits at a time. Carry over the remainders after reducing to mod 10000 if any value is more than 10000, which yield 0273, 0034 and 1 and write these numbers from right to left to get the product of 10021x10013=100340273.

$$
\begin{array}{c|c}
10021 & +21 \\
10013 & +13 \\
\hline
10034 & 273
\end{array}
$$

$$(10034)(273)=>(10034)(0273)=100340273$$

Figure 8.3: 10021x10013 = 100340273

Example 8.4: To find the product of 10002x10003, write down the numbers 10002 and 10003 vertically as written in the Figure 8.4. Then subtract the value 10000 from the numbers and write the remainders 2 and 3 at the right hand side corresponding to the respective values 10002 and 10003. Find the product of 2 and 3 and write down the resulting number 6 vertically below to the numbers 2 and 3. Now add the value 3 to 10002 or 2 to 10003; both will yield the same results and write the resulting number 10005 vertically below to the numbers 10002 and 10003. Now write these values from right to left four digits at a time. Carry over the remainders after reducing to mod 10000 if any of the values is more than 10000, which yield

0006, 0005 and 1 and write these numbers from right to left to get the product of 10002x10003=100050006

$$
\begin{array}{r|l}
10002 & +2 \\
10003 & +3 \\
\hline
10005 & 6
\end{array}
$$

$$(10005)(6)=>(10005)(0006)=100050006$$

Figure 8.4:10002x10003 = 100050006

Case II: When one of the two numbers is greater than 10000 and the other one is less than 10000

i. Let X_1 and X_2 be the two numbers to be multiplied. Write down the numbers vertically one by one as written in Example 8.5.

ii. Subtract 10000 from the numbers and write the remainders say R_1 and R_2 at the right hand side.

iii. Multiply R_1 and R_2. If the value of R_1xR_2 has less than four digits, convert it to a four digits number by adding required number of zeros at the prefix. Write down the results as done in the Example 8.5.

iv. Add R_2 with X_1 or R_1 with X_2 and write down the results as done in Example 8.5.

v. Since the product of R_1 and R_2 is negative take the complement of the last four digits and if the absolute value is more than 10000 then reduce to mod 10000 and carry over the negative value of the remainders to the next position.

vi. Write four digits at a time from right to left to get the required result.

For more details please refer the numerical examples given below:

Example 8.5: To find the product of 9998x10013, write down the numbers 9998 and 10013 vertically as written in the Figure 8.5. Then subtract the value 10000 from 9998 and 10013 and write the remainders -2 and 13 at the right hand side corresponding to the respective values 9998 and 10013. Find the product of -2 and 13 and write down the resulting number -26 vertically below to the numbers -2 and 13. Now add the value 13 to 9998 or -2 to 10013; both will yield the same result and write the resulting number 10011 vertically below to the numbers 9998 and 10013. Now as stated in step v, write these values from right to left four digits at a time. Carry over the remainders after reducing to mod 10000 if any value is more than 10000, which yield 9974, 0010 and 1 and write these numbers from right to left to get the product of 9998x10013=100109974

$$
\begin{array}{r|l}
9998 & -2 \\
10013 & +13 \\
\hline
10011 & -26 \\
\end{array}
$$

$$(10011)(-26)=>(10011)(-0026)=(10010)(0974)=100109974$$

Figure 8.5: 9998x10013 = 100109974

Example 8.6: To find the product of 9988x10006, write down the numbers 9988 and 10006 vertically as written in the Figure 8.6. Then subtract the value 10000 from the numbers 9988 and 10006 and write the remainders -12 and 6 at the right hand side corresponding to the respective values 9988 and 10006. Find the product of -12 and 6 and write down the result -72 vertically below to the numbers -12 and 6. Now add the value 6 to 9988 or -12 to 10006; both will yield the same result and write the resulting number 9994 vertically below to the numbers 9988 and 10006. Now as stated in step v, write these values from right to left four digits at a time. Carry over the remainders after reducing to mod 10000 if any value is more than 10000, which yield 9928 and 9993. Write down these numbers from right to left to get the product of 9988x10006=99939928

$$
\begin{array}{r|l}
9988 & -12 \\
10006 & +6 \\
\hline
9994 & -72
\end{array}
$$

(9994)(-72)=>(9994)(-0072)=(9993)(9928) = 99939928

Figure 8.6: 9988x10006 = 99939928

SOLVED PROBLEMS

The following are some of the solved problems given to explain the method of Multiplication based at 10000.

Problem 8.1: To find 10005x10004

10005	+5
10004	+4
10009	20

$(1009)(20)=>(10009)(0020)=100090020$

Solution: 10005x10004 = 100090020

Problem 8.2: To find 10017x10013

10017	+17
10013	+13
10010	221

$(10010)(221)=> (10010)(0221)= 100100221$

Solution: 10017x10013 = 100100221

Problem 8.3: To find 10015x10017

10015	+15
10017	+17
10032	255

$(10032)(255)=>(10032)(0255)= 100320255$

Solution: 10015x10017 = 100320255

Problem 8.4: To find 10015x9999

$$\begin{array}{c|c} 10015 & +15 \\ 9999 & -1 \\ \hline 10014 & -15 \end{array}$$

(10014)(-15)=>(10014)(-0015) =(10013)(9985)=100139985

Solution: 10015x9999 = 100139985

Problem 8.5: To find 10017x9988

$$\begin{array}{c|c} 10017 & +17 \\ 9988 & -12 \\ \hline 10005 & -204 \end{array}$$

(10005)(-204)=>(10005)(-0204)=(10004)(9796)=100049796

Solution: 10017x9988 = 100049796

Problem 8.6: To find 10009x9989

$$\begin{array}{c|c} 10009 & +9 \\ 9989 & -11 \\ \hline 9998 & -99 \end{array}$$

(9998)(-99)=>(9998)(-0099)=(9997)(9901) = 99979901

Solution: 10009x9989 = 99979901

EXERCISE PROBLEMS

The following are some of the exercise problems to be solved by using the method of Multiplication Based at 10000 to have an acquaintance with the new method of multiplication.

1.	10015x10011		16.	9995x10018
2.	10011x10012		17.	10016x9996
3.	10010x10009		18.	9998x10018
4.	10012x10016		19.	9989x10014
5.	10015x10018		20.	9997x10016
6.	10012x10015		21.	9995x10012
7.	10021x10014		22.	9987x10013
8.	10016x10013		23.	9992x10026
9.	10007x10017		24.	9988x10014
10.	10015x10015		25.	9989x10018
11.	9997x9989		26.	9997x10019
12.	9988x9995		27.	10021x10023
13.	9987x9985		28.	10024x10018
14.	9995x9992		29.	10025x10021
15.	9997x9990		30.	10015x10027

CHAPTER 9

MULTIPLICATION OF NUMBERS- BASED AT 50

In the preceding Chapters we have discussed the multiplication of two numbers based at 10, 100, 1000 and 10000. They are also illustrated with the help of several numerical examples. That is, the two numbers to be multiplied are very close to the value 10^k, where k is any positive integer. In the procedure of multiplication based at 10^k one has to write k-digits at a time. This procedure has been extended to multiplication of two numbers, which are very close to 5×10^k, where k is any positive integer. In this procedure one may come across the following two cases:

i.   Both the numbers to be multiplied are lesser (greater) than the number 5×10^k and

ii.   Out of the two numbers one may be greater than 5×10^k and other one is lesser than 5×10^k.

The other details of working procedures of these methods are explained in appropriate headings and also illustrated with the help of several numerical examples.

MULTIPLICATION BASED AT 50

The procedure of multiplication based at 50 is explained for finding the multiplication of two numbers, which are very close to either side of the value 50 and can be easily extended to other numbers like 500, 5000, etc. As mentioned above one may have the following two cases.

The steps involved are as follows:

Case I: When the numbers are greater (less) than 50.

i. Let X_1 and X_2 be the two numbers to be multiplied. Write down the numbers vertically one by one as written in Example 9.1.

ii. Subtract 50 from the numbers and write the remainders, say R_1 and R_2 at the right hand side.

iii. Multiply R_1 and R_2. If the value of $R_1 x R_2$ has less than two digits, convert it to a two digits number by adding a zero at the prefix. Write down the results as done in the Example 9.1.

iv. Add R_2 with X_1 or R_1 with X_2 and divide the result by 2 and write down the results as done in Example 9.1. If both the numbers are odd (even) write the value $(R_1+X_2)/2$ or $(R_2+X_1)/2$ as it is. Otherwise write the integral part of the above values and add 5 to the second digit from extreme right to left direction.

v. If any sum is more than 100 then reduce to mod 100 and carry over the remainders to the next position.

vi. Write the digits two at a time from right to left to get the result.

For more details please refer the numerical examples given below:

Example 9.1: To find the product of 52x54, write down the numbers 52 and 54 vertically as written in the Figure 9.1. Then subtract 50 from the numbers and write the remainders 2 and 4 at the right hand side in corresponding to the values 52 and 54. Find the product of 2 and 4 and write down the resulting number 8 vertically below to the numbers 2 and 4. Now add the value 4 to 52 or 2 to 54; both will yield the same results and write the resulting number 56 vertically below to the numbers 52 and 54. Now write these values from right to left two digits at a time. Carry over the remainders after reducing to mod 100 if any value is more than 100, which yield 08 and 28 and write these numbers from right to left to get the product of 52x54=2808

$$
\begin{array}{c|c}
52 & +2 \\
54 & +4 \\
\hline
56 & 8 \\
\end{array}
$$

$(56/2)(8)=>(28)(08)=2808$

Figure 9.1: 52x54 = 2808

Example 9.2: To find the product of 55x61, write down the numbers 55 and 61 vertically as written in the figure 9.2. Then subtract 50 from the numbers and write the remainders 5 and 11 at the right hand side in corresponding to the values 55 and 61. Find the product of 5 and 11 and write down the resulting number 55 vertically below to the numbers 5 and 11. Now add the value 5 to 61 or 11 to 55; both will yield the same results and write the resulting number 66 vertically below to the numbers 55 and 61. Write these values from right to left two digits at a time and carry over the remainders after reducing to mod 100 if any value is more than 100, which yield 55 and 33 and write these numbers from right to left to get the product of 55x61=3355

$$
\begin{array}{c|c}
55 & +5 \\
61 & +11 \\
\hline
66 & 55 \\
\end{array}
$$

$(66/2)(55)=>(33)(55)=3355$

Figure 9.2:55x61 = 3355

Example 9.3: To find the product of 55x62, write down the numbers 55 and 62 vertically as written in the Figure 9.3. Then subtract 50 from the numbers and write the remainders 5 and 12 at the right hand side in corresponding to the values 55 and 62. Find the product of 5 and 12 and write down the resulting number 60 vertically below to the numbers 5 and 12. Now add the value 5 to 62 or 12 to 55; both will yield the same results and write the resulting number 67 vertically below to the numbers 55 and 62. Now write these values from right to left two digits at a time. Carry over the remainders after reducing to mod 100 if any value is more than 100, which yield 10 and 34 and write these numbers from right to left to get the product of 55x62=3355

$$
\begin{array}{c|c}
55 & +5 \\
62 & +12 \\
\hline
67 & 60
\end{array}
$$

$$(67/2)(60) => (33.5)(60) = (33)(110) = 3410$$

Figure 9.3: 55x62 = 3410

Note: As stated in step iv if both the numbers are odd (even) write the value $(R_1+X_2)/2$ or $(R_2+X_1)/2$ as it is. Otherwise write the integral part of the above values and add 5 to the second digit from extreme right to left direction.

Case II: When one of the two numbers is greater than 50 and the other one is less than 50

i. Let X_1 and X_2 be the two numbers to be multiplied. Write down the numbers vertically one by one as written in Example 9.4.

ii. Subtract 50 from the numbers and write the remainders say R_1 and R_2 at the right hand side.

iii. Multiply R_1 and R_2. If the value of R_1xR_2 has less than two digits, convert it to a two digits number by adding a zero at the prefix. Write down the results as done in the Example 9.4.

iv. Add R_2 with X_1 or R_1 with X_2 and divide by 2 and write down the results as done in Example 9.4. If both the numbers are odd (even) write the value $(R_1+X_2)/2$ or $(R_2+X_1)/2$ as it is. Otherwise write the integral part of the above values and add 5 to the second digit from extreme right to left direction.

v. Since the product of R_1 and R_2 is negative take the complement of the last two digits and if the absolute value is more than 100 then reduce to mod 100 and carry over the negative value of remainders to the next position.

vi. Write two digits at a time from right to left to get the result.

For more details please refer the numerical examples given below:

Example 9.4: To find the product of 45x53, write down the numbers 45 and 53 vertically as written in the Figure 9.4. Then subtract 50 from the numbers and write the remainders -5 and 3 at the right hand side in corresponding to the values 45 and 53. Find the product of -5 and 3 and write down the resulting number -15 vertically below to the numbers -5 and 3. Now add the value 3 to 45 or -5 to 53; both will yield the same results and write the resulting number 48 vertically below to the numbers 45 and 53. As stated in step v, write these values from right to left two digits at a time. Carry over the remainders after reducing to mod 100 if any value is more than 100, which yield 85 and 23 and write these numbers from right to left to get the product of 45x53=2385

$$
\begin{array}{c|c}
45 & -5 \\
53 & +3 \\
\hline
48 & -15 \\
\end{array}
$$

$(48/2)(-15)=>(24)(-15)=(23)(85)=2385$

Figure 9.4: 45x53 = 2385

Example 9.5: To find the product of 45x62, write down the numbers 45 and 62 vertically as written in the Figure 9.5. Then subtract 50 from the numbers and write the remainders -5 and 12 at the right hand side in corresponding to the values 45 and 62. Find the product of -5 and 12 and write down the resulting number -60 vertically below to the numbers -5 and 12. Now add the value 12 to 45 or -5 to 62; both will yield the same results and write the resulting number vertically below to the numbers 45 and 62. As stated in step v, write these values from right to left two digits at a time. Carry over the remainders after reducing to mod 100 if any value is more than 100, which yield 90 and 27 and write these numbers from right to left to get the product of 45x62=2790

$$
\begin{array}{c|c}
45 & -5 \\
62 & +12 \\
\hline
57 & -60 \\
\end{array}
$$

$$(57/2)(-60)=>(28.5)(-60)=(27.5)(40)=2790$$

Figure 9.5: 45x62 = 2790

SOLVED PROBLEMS

The following are some of the solved problems given to explain the method of Multiplication Based at 50:

Problem 9.1: To find 58x54

58	+8
54	+4
62	32

$(62/2)(32)=>(31)(32)=3132$

Solution: 58x54 = 3132

Problem 9.2: To find 57x53

57	+7
53	+3
60	21

$(60/2)(21)=>(30)(21)=3021$

Solution: 57x53 = 3021

Problem 9.3: To find 65x57

65	+15
57	+7
72	105

$(72/2)(105)=>(36)(105)=(36+1)(05) = 3705$

Solution: 65x57 = 3705

Problem 9.4: To find 55x49

$$
\begin{array}{c|c}
55 & +5 \\
49 & -1 \\
\hline
54 & -5
\end{array}
$$

$(54/2)(-5)=>(27)(-05)=(26)(95) = 2695$

Solution: 55x49 = 2695

Problem 9.5: To find 57x48

$$
\begin{array}{c|c}
57 & +7 \\
48 & -2 \\
\hline
55 & -14
\end{array}
$$

$(55/2)(-14)=>(27.5)(-14)=(26.5)(86)=2736$

Solution: 57x48 = 2736

Problem 9.6: To find 59x49

$$
\begin{array}{c|c}
59 & +9 \\
49 & -1 \\
\hline
58 & -9
\end{array}
$$

$(58/2)(-09)=>(29)(-09)=(28)(91) = 2891$

Solution: 59x49 = 2891

Problem 9.7: To find 45x49

$$\begin{array}{c|c} 45 & -5 \\ 49 & -1 \\ \hline 44 & +5 \end{array}$$

$$(44/2)(5)=>(22)(05)=2205$$

Solution: 45x49 = 2205

Problem 9.8: To find 47x48

$$\begin{array}{c|c} 47 & -3 \\ 48 & -2 \\ \hline 45 & +6 \end{array}$$

$$(45/2)(6)=>(22.5)(06)=(22)(56)= 2256$$

Solution: 47x48 = 2256

Problem 9.9: To find 49x41

$$\begin{array}{c|c} 49 & -1 \\ 41 & -9 \\ \hline 40 & +9 \end{array}$$

$$(40/2)(9)=>(20)(09)= 2009$$

Solution: 49x41 = 2009

EXERCISE PROBLEMS

The following are some of the exercise problems to be solved by using the method of Multiplication Based at 50 to have an acquaintance with the new method of multiplication.

1. 55x61	16. 45x58
2. 61x62	17. 46x59
3. 60x59	18. 48x58
4. 62x56	19. 49x64
5. 55x57	20. 44x56
6. 62x65	21. 43x63
7. 52x64	22. 37x63
8. 51x51	23. 39x56
9. 57x63	24. 48x64
10. 65x65	25. 38x58
11. 49x48	26. 55x66
12. 48x45	27. 61x63
13. 48x48	28. 54x68
14. 39x49	29. 55x61
15. 47x43	30. 65x51

CHAPTER 10

MULTIPLICATION OF NUMBERS- BASED AT 500

In the preceding Chapters we have discussed the multiplication of two numbers based at 10^k, for any positive integer k. However for the sake of convenience we have discussed explicitly for the multiplication of two numbers when k= 1, 2, 3 and 4. They are also illustrated with the help of several numerical examples. This procedure has been extended to multiplication of two numbers, which are very close to 50 and is explained with several numerical problems in Chapter 10. In this Chapter this method has been extended for the multiplication of numbers very close to 500. In this procedure one may come across the following two cases:

i.   Both the numbers to be multiplied are lesser (greater) than the number 500 and

ii. Out of the two numbers one may be greater than 500 and other one is lesser than 500.

The other details of working procedures of these methods are explained in appropriate headings and also illustrated with the help of several numerical examples.

MULTIPLICATION BASED AT 500

The procedure of multiplication based at 500 is explained for finding the multiplication of two numbers, which are very close to either side of the value 500. As mentioned above one may have the following two cases.

The steps involved are as follows:

Case I: When the numbers are greater (less) than 500.

i. Let X_1 and X_2 be the two numbers to be multiplied. Write down the numbers vertically one by one as written in Example 10.1.

ii. Subtract 500 from the numbers and write the remainders, say R_1 and R_2 at the right hand side.

iii. Multiply R_1 and R_2. If the value of $R_1 x R_2$ has less than three digits, convert it to a three digits number by adding required number of zeros at the prefix. Write down the results as done in the Example 10.1.

iv. Add R_2 with X_1 or R_1 with X_2 and divide the result by 2 and write down the results as done in Example 10.1. If both the numbers are odd (even) write the value $(R_1+X_2)/2$ or $(R_2+X_1)/2$ as it is. Otherwise write the integral part of the above values and add 5 to the third digit from extreme right to left direction.

v. If any sum is more than 1000 then reduce to mod 1000 and carry over the remainders to the next position.

vi. Write the digits three at a time from right to left to get the result.

For more details please refer the numerical examples given below:

Example 10.1: To find the product of 502x504, write down the numbers 502 and 504 vertically as written in the Figure 10.1. Then subtract 500 from the numbers and write the remainders 2 and 4 at the right hand side in corresponding to the values 502 and 504. Find the product of 2 and 4 and write down the resulting number 8 vertically below to the numbers 2 and 4. Now add the value 4 to 502 or 2 to 504; both will yield the same results and write the resulting number 506 vertically below to the numbers 502 and 504. Now write these values from right to left three digits at a time. Carry over the remainders after reducing to mod 1000 if any value is more than 1000, which yield 008 and 253 and write these numbers from right to left one may get the product of 502x504=253008

502	+2
504	+4
506	8

$$(506/2)(8)=>(253)(008)=253008$$

Figure 10.1: 502 x504 = 253008

Example 10.2: To find the product of 505x511, write down the numbers 505 and 511 vertically as written in the Figure 10.2. Then subtract 500 from the numbers and write the remainders 5 and 11 at the right hand side in corresponding to the values 505 and 511. Find the product of 5 and 11 and write down the resulting number 55 vertically below to the numbers 5 and 11. Now add the value 5 to 511 or 11 to 505; both will yield the same results and write the resulting number 516 vertically below to the numbers 505 and 511. Write these values from right to left three digits at a time. Carry over the remainders after reducing to mod 1000 if any value is more than 1000, which yield 055 and 258 and write these numbers from right to left to get the product of 505x511=258055

505	+5
511	+11
516	55

$$(516/2)(55)=>(258)(055)=258055$$

Figure 10.2:505x511 = 258055

Example 10.3: To find the product of 515x512, write down the numbers 515 and 512 vertically as written in the Figure 10.3. Then subtract 500 from the numbers and write the remainders 15 and 12 at the right hand side in corresponding to the values 515 and 512. Find the product of 15 and 12 and write down the resulting number 180 vertically below to the numbers 15 and 12. Now add the value 15 to 512 or 12 to 515; both will yield the same results and write the resulting number 527 vertically below to the numbers 515 and 512. Now write these values from right to left three digits at a time. Carry over the remainders after reducing to mod 1000 if any value is more than 1000, which yield 680 and 253 and write these numbers from right to left one may get the product of 515x512=263680

$$
\begin{array}{c|c}
515 & +15 \\
512 & +12 \\
\hline
527 & 180 \\
\end{array}
$$

$$(527/2)(180)=>(263.5)(180)=(263)(680)=263680$$

Figure 10.3: 515x512 = 263680

Note: As stated in step iv if both the numbers are odd (even) write the value $(R_1+X_2)/2$ or $(R_2+X_1)/2$ as it is. Otherwise write the integral part of the above values and add 5 to the third digit from extreme right to left direction.

Case II: When one of the two numbers is greater than 500 and the other one is less than 500

i.   Let X_1 and X_2 be the two numbers to be multiplied. Write down the numbers vertically one by one as written in Example 10.4.

ii. Subtract 500 from the numbers and write the remainders say R_1 and R_2 at the right hand side.

iii. Multiply R_1 and R_2. If the value of R_1xR_2 has less than three digits, convert it to a three digits number by adding required number of zeros at the prefix. Write down the results as done in the Example 10.4.

iv. Add R_2 with X_1 or R_1 with X_2 and divide by 2 and write down the results as done in Example 10.4. If both the numbers are odd (even) write the value $(R_1+X_2)/2$ or $(R_2+X_1)/2$ as it is. Otherwise write the integral part of the above values and add 5 to the third digit from extreme right to left direction.

v. Since the product of R_1 and R_2 is negative take the complement of the last three digits and if the absolute value is more than 1000 then reduce to mod 1000 and carry over the negative value of remainders to the next position.

vi. Write three digits at a time from right to left to get the result.

For more details please refer the numerical examples given below:

Example 10.4: To find the product of 495x503, write down the numbers 495 and 503 vertically as written in the Figure 10.4. Then subtract 500 from the numbers and write the remainders -5 and 3 at the right hand side in corresponding to the values 495 and 503. Find the product of -5 and 3 and write down the resulting number -15 vertically below to the numbers -5 and 3. Now add the value 3 to 495 or -5 to 503; both will yield the same results and write the resulting number 498 vertically below to the numbers 495 and 503. As stated in step v, write these values from right to left three digits at a time. Carry over the remainders after reducing to mod 1000 if any value is more than 1000, which yield 985 and 248 and write these numbers from right to left to get the product of 495x503=248985

$$
\begin{array}{c|c}
495 & -5 \\
503 & +3 \\
\hline
498 & -15
\end{array}
$$

$$(498/2)(-15)=>(249)(-015)=(248)(985)=248985$$

Figure 10.4: 495x503 = 248985

Example 10.5: To find the product of 495x512, write down the numbers 495 and 512 vertically as written in the Figure 10.5. Then subtract 500 from the numbers and write the remainders -5 and 12 at the right hand side in corresponding to the values 495 and 512. Find the product of -5 and 12 and write down the resulting number -60 vertically below to the numbers -5 and 12. Now add the value 12 to 495 or -5 to 512; both will yield the same results and write the resulting number 507 vertically below to the numbers 495 and 512. As stated in step v, write these values from right to left three digits at a time. Carry over the remainders after reducing to mod 1000 if any value is more than 1000, which yield 440 and 253 and write these numbers from right to left to get the product of 495x512=253440

$$
\begin{array}{r|l}
495 & -5 \\
512 & +12 \\
\hline
507 & -60 \\
\end{array}
$$

(507/2)(-60)=>(253.5)(-060)=(252.5)(940)=253440

Figure 10.5: 495x512 = 253440

SOLVED PROBLEMS

The following are some of the solved problems given to explain the method of Multiplication Based at 500.

Problem 10.1: To find 508x514

$$
\begin{array}{c|c}
508 & +8 \\
514 & +14 \\
\hline
522 & 112
\end{array}
$$

$(522/2)(112)=>(261)(112)=261112$

Solution: 508x514 = 261112

Problem 10.2: To find 507x513

$$
\begin{array}{c|c}
507 & +7 \\
513 & +13 \\
\hline
520 & 91
\end{array}
$$

$(520/2)(91)=>(260)(091)=260091$

Solution: 507x513 = 260091

Problem 10.3: To find 515x507

$$
\begin{array}{c|c}
515 & +15 \\
507 & +7 \\
\hline
522 & 105
\end{array}
$$

$(522/2)(105)=>(261)(105)= 261105$

Solution: 515x507 = 261105

Problem 10.4: To find 515x499

$$
\begin{array}{r|l}
515 & +15 \\
499 & -1 \\
\hline
514 & -15
\end{array}
$$

(514/2)(-15)=>(257)(-015)=(256)(985) = 256985

Solution: 515x499 = 256985

Problem 10.5: To find 517x498

$$
\begin{array}{r|l}
517 & +17 \\
498 & -2 \\
\hline
515 & -34
\end{array}
$$

(515/2)(-34)=>(257.5)(-034)=(256.5)(966)=257466

Solution: 517x498 = 257466

Problem 10.6: To find 509x489

$$
\begin{array}{r|l}
509 & +9 \\
489 & -11 \\
\hline
498 & -99
\end{array}
$$

(498/2)(-99)=>(249)(-099)=(248)(901) = 248901

Solution: 509x489 = 248901

Problem 10.7: To find 495x489

$$
\begin{array}{r|l}
495 & -5 \\
489 & -11 \\
\hline
484 & +55
\end{array}
$$

(484/2)(55)=>(242)(055)=242055

Solution: 495x489 = 242055

Problem 10.8: To find 493x498

$$
\begin{array}{c|c}
493 & -7 \\
498 & -2 \\
\hline
491 & +14
\end{array}
$$

$$(491/2)(14) => (245.5)(014) = (245)(514) = 245514$$

Solution: 493x498 = 245514

Problem 10.9: To find 499x491

$$
\begin{array}{c|c}
499 & -1 \\
491 & -9 \\
\hline
490 & +9
\end{array}
$$

$$(490/2)(9) => (245)(009) = 245009$$

Solution: 499x491 = 245009

EXERCISE PROBLEMS

The following are some of the exercise problems to be solved by using the method of Multiplication Based at 500 to have an acquaintance with the new method of multiplication.

1. 505x511	16. 495x508
2. 511x512	17. 496x519
3. 510x509	18. 498x518
4. 512x506	19. 494x514
5. 505x517	20. 494x516
6. 502x515	21. 493x513
7. 512x514	22. 487x513
8. 513x517	23. 489x516
9. 509x513	24. 488x514
10. 519x515	25. 498x518
11. 491x498	26. 525x516
12. 478x495	27. 521x513
13. 489x498	28. 524x518
14. 493x499	29. 525x511
15. 497x493	30. 515x523

CHAPTER 11

MULTIPLICATION OF NUMBERS- BASED AT 5000

In the preceding Chapters we have discussed the multiplication of two numbers based at $5*10^k$, for any positive integer k. However for the sake of convenience we have discussed explicitly for the multiplication of two numbers when k= 1 and 2. That is the procedure for the multiplication of two numbers, which are very close to 50 and 500. They are also explained with the help of several numerical problems in Chapter 9 and 10 respectively. In this Chapter this method has been extended for the multiplication of numbers very close to 5000. As discussed earlier, in this procedure one may come across the following two cases:

i.   Both the numbers to be multiplied are less (greater) than the number 5000 and

ii. One of the two numbers is greater than 5000 and the other one is less than 5000.

The other details of working procedures of these methods are explained in appropriate headings and also illustrated with the help of several numerical examples.

MULTIPLICATION BASED AT 5000

The procedure of multiplication based at 5000 is explained for finding the multiplication of two numbers, which are very close to either side of 5000. As mentioned above one may have the following two cases.

The steps involved are as follows:

Case I: When both numbers are greater (lesser) than 5000

i. Let X_1 and X_2 be the two numbers to be multiplied. Write down the numbers vertically one by one as written in Example 11.1.

ii. Subtract 5000 from the numbers and write the remainders, say R_1 and R_2 at the right hand side.

iii. Multiply R_1 and R_2. If the value of $R_1 x R_2$ has less than four digits, convert it to a four digits number by adding required number of zeros at the prefix. Write down the results as done in the Example 11.1.

iv. Add R_2 with X_1 or R_1 with X_2 and divide the result by 2 and write down the results as done in Example 11.1. If both the numbers are odd (even) write the value $(R_1+X_2)/2$ or $(R_2+X_1)/2$ as it is. Otherwise write the integral part of the above values and add 5 to the fourth digit from extreme right to left direction.

v. If any sum is more than 10000 then reduce to mod 10000 and carry over the remainders to the next position.

vi. Write the digits four at a time from right to left to get the result.

For more details please refer the numerical examples given below:

Example 11.1: To find the product of 5002x5004, write down the numbers 5002 and 5004 vertically as written in the Figure 11.1. Then subtract 5000 from the numbers and write the remainders 2 and 4 at the right hand side in corresponding to the values 5002 and 5004. Find the product of 2 and 4 and write down the resulting number 8 vertically below to the numbers 2 and 4. Now add the value 4 to 5002 or 2 to 5004; both will yield the same results and write the resulting number 5006 vertically below to the numbers 5002 and 5004. Now write these values from right to left four digits at a time.

Carry over the remainders after reducing to mod 10000 if any value is more than 10000, which yield 0008 and 2503 and write these numbers from right to left to get the product of 5002x5004=25030008

$$\begin{array}{c|c} 5002 & +2 \\ 5004 & +4 \\ \hline 5006 & 8 \end{array}$$

$(5006/2)(8)=>(2503)(0008)=25030008$

Figure 11.1: 5002 x5004 = 25030008

Example 11.2: To find the product of 5005x5011, write down the numbers 5005 and 5011 vertically as written in the Figure 11.2. Then subtract 5000 from the numbers and write the remainders 5 and 11 at the right hand side in corresponding to the values 5005 and 5011. Find the product of 5 and 11 and write down the resulting number 55 vertically below to the numbers 5 and 11. Now add the value 5 to 5011 or 11 to 5005; both will yield the same results and write the resulting number 5016 vertically below to the numbers 5005 and 5011. Write these values from right to left four digits at a time. Carry over the remainders after reducing to mod 10000 if any value is more than 10000, which yield 0055 and 2508 and write these numbers from right to left to get the product of 5005x5011=25080055

$$\begin{array}{c|c} 5005 & +5 \\ 5011 & +11 \\ \hline 5016 & 55 \end{array}$$

$(5016/2)(55)=>(2508)(0055)=25080055$

Figure 11.2:5005x5011 = 25080055

Example 11.3: To find the product of 5015x5012, write down the numbers 5015 and 5012 vertically as written in the Figure 11.3. Then subtract 5000 from the numbers and write the remainders 15 and 12 at the right hand side in corresponding to the values 5015 and 5012. Find the product of 15 and 12 and write down the resulting number 180 vertically below to the numbers 15 and 12. Now add the value 15 to 5012 or 12 to 5015; both will yield the same results and write the resulting number 5027 vertically below to the numbers 5015 and 5012. Now write these values from right to left four digits at a time. Carry over the remainders after reducing to mod 10000 if any value is more than 10000, which yield 5180 and 2513 and write these numbers from right to left one may get the product of 5015x5012= 2 5135180

$$
\begin{array}{c|c}
5015 & +15 \\[2mm]
5012 & +12 \\[2mm]
\hline
5027 & 180 \\
\end{array}
$$

$$(5027/2)(180)=>(2513.5)(0180)=(2513)(5180)=25135180$$

Figure 11.3: 5015x5012 = 25135180

Note: As stated in step iv if both the numbers are odd (even) write the value $(R_1+X_2)/2$ or $(R_2+X_1)/2$ as it is. Otherwise write the integral part of the above values and add 5 to the fourth digit from extreme right to left direction.

Case II: When one of the two numbers is greater than 5000 and the other one is less than 5000

i. Let X_1 and X_2 be the two numbers to be multiplied. Write down the numbers vertically one by one as written in Example 11.4.

ii. Subtract 5000 from the numbers and write the remainders say R_1 and R_2 at the right hand side.

iii. Multiply R_1 and R_2. If the value of $R_1 x R_2$ has less than four digits, convert it to a four digits number by adding required number of zeros at the prefix.

iv. Add R_2 with X_1 or R_1 with X_2 and divide by 2 and write down the results as done in Example 11.4. If both the numbers are odd (even) write the value $(R_1+X_2)/2$ or $(R_2+X_1)/2$ as it is. Otherwise write the integral part of the above values and add 5 to the second digit from extreme right to left direction.

v. Since the product of R_1 and R_2 is negative take the complement of the last four digits and if the absolute value is more than 10000 then reduce to mod 10000 and carry over the negative value of remainders to the next position.

vi. Write four digits at a time from right to left to get the result.

For more details please refer the numerical examples given below:

Example 11.4: To find the product of 4995x5003, write down the numbers 4995 and 5003 vertically as written in the Figure 11.4. Then subtract 5000 from the numbers and write the remainders -5 and 3 at the right hand side in corresponding to the values 4995 and 5003. Find the product of -5 and 3 and write down the resulting number -15 vertically below to the numbers -5 and 3. Now add the value 3 to 4995 or -5 to 5003; both will yield the same results and write the resulting number 4998 vertically below to the numbers 4995 and 5003. As stated in step v, write these values from right to left four digits at a time. Carry over the remainders after reducing to mod 10000 if any value is more than 10000, which yield 9985 and 2498 and write these numbers from right to left to get the product of 4995x5003= 24989985

$$
\begin{array}{c|c}
4995 & -5 \\
5003 & +3 \\
\hline
4998 & -15
\end{array}
$$

$(4998/2)(-15)=>(2499)(-0015)=(2498)(9985)=24989985$

Figure 11.4: 4995x5003 = 24989985

Example 11.5: To find the product of 4995x5012, write down the numbers 4995 and 5012 vertically as written in the Figure 11.5. Then subtract 5000 from the numbers and write the remainders -5 and 12 at the right hand side in corresponding to the values 4995 and 5012. Find the product of -5 and 12 and write down the resulting number -60 vertically below to the numbers -5 and 12. Now add the value 12 to 4995 or -5 to 5012; both will yield the same results and write the resulting number 5007 vertically below to the numbers 4995 and 5012. As stated in step v, write these values from right to left four digits at a time. Carry over the remainders after reducing to mod 10000 if any value is more than 10000, which yield 4940 and 2503 and write these numbers from right to left to get the product of 4995x5012= 253440

$$
\begin{array}{c|c}
4995 & -5 \\
5012 & +12 \\
\hline
5007 & -60
\end{array}
$$

$(5007/2)(-60)=>(2503.5)(-0060)= (2502.5)(9940)=25034940$

Figure 11.5: 4995x5012 = 25034940

SOLVED PROBLEMS

The following are some of the solved problems given to explain the method of Multiplication Based at 5000.

Problem 11.1: To find 5008x5014

5008	+8
5014	+14
5022	112

$$(5022/2)(112) => (2511)(0112) = 25110112$$

Solution: 5008x5014 = 25110112

Problem 11.2: To find 5007x5013

5007	+7
5013	+13
5020	91

$$(5020/2)(91) => (2510)(0091) = 25100091$$

Solution: 5007x5013 = 25100091

Problem 11.3: To find 5015x5007

5015	+15
5007	+7
5022	105

$$(5022/2)(105) => (2511)(0105) = 25110105$$

Solution: 5015x5007 = 25110105

Problem 11.4: To find 5015x4999

$$
\begin{array}{r|l}
5015 & +15 \\
4999 & -1 \\
\hline
5014 & -15 \\
\end{array}
$$

(5014/2)(-15)=>(2507)(-0015)=(2506)(9985)=25069985

Solution: 5015x4999 = 25069985

Problem 11.5: To find 5017x4998

$$
\begin{array}{r|l}
5017 & +17 \\
4998 & -2 \\
\hline
5015 & -34 \\
\end{array}
$$

(5015/2)(-34)=>(2507.5)(-0034)=(2506.5)(9966)=25074966

Solution: 5017x4998 = 25074966

Problem 11.6: To find 5009x4989

$$
\begin{array}{r|l}
5009 & +9 \\
4989 & -11 \\
\hline
4980 & -99 \\
\end{array}
$$

(4980/2)(-99)=>(2490)(-0099)=(2489)(9901)=24899901

Solution: 5009x4989 = 24899901

Problem 11.7: To find 4995x4989

$$
\begin{array}{r|l}
4995 & -5 \\
4989 & -11 \\
\hline
4984 & +55 \\
\end{array}
$$

(4984/2)(55)=>(2492)(0055)=24920055

Solution: 4995x4989 = 24920055

Problem 11.8: To find 4993x4998

4993	-7
4998	-2
4991	+14

(4991/2)(14)=>(2495.5)(0014)=(2495)(5014)=24955014

Solution: 4993x4998 = 24955014

Problem 11.9: To find 4999x4991

4999	-1
4991	-9
4990	+9

(4990/2)(9)=>(2495)(0009)= 24950009

Solution: 4999x4991 = 24950009

EXERCISE PROBLEMS

The following are some of the exercise problems to be solved by using the method of Multiplication Based at 5000 to have an acquaintance with the new method of multiplication.

1. 5005x5011	16. 4995x5008
2. 5011x5012	17. 4996x5019
3. 5010x5009	18. 4998x5018
4. 5012x5006	19. 4994x5014
5. 5005x5017	20. 4994x5016
6. 5002x5015	21. 4993x5013
7. 5012x5014	22. 4987x5013
8. 5013x5017	23. 4989x5016
9. 5009x5013	24. 4988x5014
10. 5019x5015	25. 4998x5018
11. 4991x4998	26. 5025x5016
12. 4978x4995	27. 5021x5013
13. 4989x4998	28. 5024x5018
14. 4993x4999	29. 5025x5011
15. 4997x4993	30. 5015x5023

CHAPTER 12

MULTIPLICATION OF SPECIAL NUMBERS

In the preceding Chapters we have discussed the multiplication based at $5*10^k$ and 10^k, where k is any positive integer. The two numbers are very close to either side of the value $5*10^k$ and 10^k. The procedure is true for any integer value of k. For the sake of convenience of the readers the procedure has been discussed explicitly for the values of k=1, 2, 3 and 4. That is the multiplication of two numbers based at 10, 50, 100, 500, 1000, 5000 and 10000 and is also illustrated with the help of several numerical examples. In the procedure of multiplication based at 10, one has to write one digit at a time; in the multiplication based at 100 one has to write two digits at a time whereas in the case of multiplication based at $5*10^k$ or 10^k one has to write k-digits at a time. In this chapter we are planning to discuss some methods of finding the product some type of specialized numbers.

The other details of working procedures of these methods are explained in appropriate headings and also illustrated with the help of several numerical examples.

MULTIPLICATION OF NUMBERS ENDING WITH 5

The procedure of multiplication of two numbers ending with 5 is explained here. The method is also illustrated with the help of several numerical problems. In this procedure one may come across the following two cases:

i.   When both the numbers are even (odd)

ii.   When one of the two numbers is even and the other one is odd

The steps involved are as follows:

Case I: When both the numbers are even (odd)

i.   Let X_1 and X_2 be the two numbers to be multiplied. Write down the numbers vertically one by one as written in Example 12.1

ii.   Remove the rightmost digit 5 from the numbers and write the remainders say R_1 and R_2 at the left hand side and 5 at the right hand side.

iii. Determine the product R_1*R_2 and average $(R_1+R_2)/2$.

iv.   Add the product with the average. That is to find $[(R_2*R_1)+(R_1+R_2)/2]$ and write down the result as done in Example 12.1.

v.   Write down the digits within the brackets from right to left to get the result.

For more details please refer the numerical examples given below:

Example 12.1: To find the product of 25x45, write down the numbers 25 and 45 vertically as written in the Figure 12.1. Find the product of 5 and 5; product of 2 and 4; average of 2 and 4 and write down the results as done in the Figure 12.1. Write these numbers from right to left to get the product of 25x45=1125

$$
\begin{array}{c|c}
2 & 5 \\
4 & 5 \\
\hline
8 & 25 \\
\end{array}
$$

$$(8+(2+4)/2)(25) => (8+3)(25)=1125$$

Figure 12.1:25x45 = 1125

Example 12.2: To find the product of 25x65, write down the numbers 25 and 65 vertically as written in the Figure 12.2. Find the product of 5 and 5; product of 2 and 6; average of 2 and 6 and write down the results as done in the Figure 12.2. Write these numbers from right to left to get the product of 25x65=1625

<table>
<tr><td>2</td><td>5</td></tr>
<tr><td>6</td><td>5</td></tr>
<tr><td>12</td><td>25</td></tr>
</table>

(12+(2+6)/2)(25)=> (12+4)(25)=1625

Figure 12.2:25x65 = 1625

Example 12.3: To find the product of 15x35, write down the numbers 15 and 35 vertically as written in the Figure 12.3. Find the product of 5 and 5; product of 1 and 3; average of 1 and 3 and write down the results as done in the Figure 12.3. Write these numbers from right to left to get the product of 15x35=525

<table>
<tr><td>1</td><td>5</td></tr>
<tr><td>3</td><td>5</td></tr>
<tr><td>3</td><td>25</td></tr>
</table>

(3+(1+3)/2)(25)=> (3+2)(25)=525

Figure 12.3:15x35 = 525

Example 12.4: To find the product of 35x55, write down the numbers 35 and 55 vertically as written in the Figure 12.4. Find the product of 5 and 5; product of 3 and 5; average of 3 and 5 and write down the results as done in the Figure 12.4. Write these numbers from right to left to get the product of 35x55=1925

$$
\begin{array}{c|c}
3 & 5 \\
5 & 5 \\
\hline
15 & 25
\end{array}
$$

$(15+(3+5)/2)(25) => (15+4)(25)=1925$

Figure 12.4:35x55 = 1925

Example 12.5: To find the product of 115x135, write down the numbers 115 and 135 vertically as written in the Figure 12.5. Find the product of 5 and 5; product of 11 and 13; average of 11 and 13 and write down the results as done in the Figure 12.5. Write these numbers from right to left to get the product of 115x135=15525

$$
\begin{array}{c|c}
11 & 5 \\
13 & 5 \\
\hline
143 & 25
\end{array}
$$

$(143+(11+13)/2)(25) => (143+12)(25)=15525$

Figure 12.5:115x135 = 15525

Case II: When one of the two numbers is even and the other one is odd

i. Let X_1 and X_2 be the two numbers to be multiplied. Write down the numbers vertically one by one as written in Example 12.6.

ii. Remove the rightmost digit 5 from the numbers and write the remainders say R_1 and R_2 at the left hand side and 5 at the right hand side.

iii. Determine the product R_1*R_2 and average $(R_1+R_2)/2$.

iv. Add the product with the average. That is find $[(R_2*R_1)+(R_1+R_2)/2]$. Write down the integral part of the above values and add 5 to the second digit from extreme right to left direction as done in Example 12.6.

v. Write down the digits within the brackets from right to left to get the result.

For more details please refer the numerical examples given below:

Note: If both the numbers are odd (even) write the value as it is. Otherwise write the integral part of the above values and add 5 to the second digit from extreme right to left direction.

Example 12.6: To find the product of 25x35, write down the numbers 25 and 35 vertically as written in the Figure 12.6. Find the product of 5 and 5; product of 2 and 3; average of 2 and 3 and write down the results as done in the Figure 12.6. Write these numbers from right to left to get the product of 25x35=875

2	5
3	5
6	75

$(6+(2+3)/2)(25) => (8.5)(25) = 875$

Figure 12.6:25x35 = 875

Example 12.7: To find the product of 25x55, write down the numbers 25 and 55 vertically as written in the Figure 12.7. Find the product of 5 and 5; product of 2 and 5; average of 2 and 5 and write down the results as done in the Figure 12.7. Write these numbers from right to left to get the product of 25x55=1375

2	5
5	5
10	75

$(10+(2+5)/2)(25) => (13.5)(25) = 1375$

Figure 12.7:25x55 = 1375

Example 12.8: To find the product of 35x65, write down the numbers 35 and 65 vertically as written in the Figure 12.8. Find the product of 5 and 5; product of 3 and 6; average of 3 and 6 and write down the results as done in the Figure 12.8. Write these numbers from right to left to get the product of 35x65=2275

$$
\begin{array}{c|c}
3 & 5 \\
6 & 5 \\
\hline
18 & 25
\end{array}
$$

$$(18+(3+6)/2)(25) => (22.5)(25)=2275$$

Figure 12.8:35x65 = 2275

Example 12.9: To find the product of 15x85, write down the numbers 15 and 85 vertically as written in the Figure 12.9. Find the product of 5 and 5; product of 1 and 8; average of 1 and 8 and write down the results as done in the Figure 12.9. Write these numbers from right to left to get the product of 15x85=1275

$$
\begin{array}{c|c}
1 & 5 \\
8 & 5 \\
\hline
8 & 25
\end{array}
$$

$$(8+(1+8)/2)(25) => (12.5)(25)=1275$$

Figure 12.9:15x85 = 1275

Example 12.10: To find the product of 125x55, write down the numbers 125 and 55 vertically as written in the Figure 12.10. Find the product of 5 and 5; product of 12 and 5; average of 12 and 5 and write down the results as done in the Figure 12.10. Write these numbers from right to left to get the product of 125x55=1925

$$
\begin{array}{c|c}
12 & 5 \\
5 & 5 \\
\hline
60 & 25
\end{array}
$$

$(60+(12+5)/2)(25)\Rightarrow (68.5)(25)=6875$

Figure 12.10:125x55 = 6875

Example 12.11: To find the product of 125x135, write down the numbers 125 and 135 vertically as written in the Figure 12.11. Find the product of 5 and 5; product of 12 and 13; average of 12 and 13 and write down the results as done in the Figure 12.11. Write these numbers from right to left to get the product of 125x135=15525

$$
\begin{array}{c|c}
12 & 5 \\
13 & 5 \\
\hline
156 & 25
\end{array}
$$

$(156+(12+13)/2)(25)\Rightarrow (168.5)(25)=16875$

Figure 12.11:125x135 = 16875

Solved Problems

The following are some of the solved problems given to explain the method of Multiplication of numbers ending with 5.

Problem 12.1: To find 45x65

4	5
6	5
24	25

$$(24+(4+6)/2)(25) => (29)(25) = 2925$$

Solution: 45x65 = 2925

Problem 12.2: To find 55x75

5	5
7	5
35	25

$$(35+(5+7)/2)(25) => (41)(25) = 4125$$

Solution: 55x75 = 4125

Problem 12.3: To find 35x85

3	5
8	5
24	25

$$(24+(3+8)/2)(25) => (29.5)(25) = 2975$$

Solution: 35x85 = 2975

Problem 12.4: To find 25x95

$$\begin{array}{c|c} 2 & 5 \\ 9 & 5 \\ \hline 18 & 25 \end{array}$$

$$(18+(2+9)/2)(25)=>(23.5)(25)=2375$$

Solution: 25x95 = 2375

Problem 12.5: To find 125x155

$$\begin{array}{c|c} 12 & 5 \\ 15 & 5 \\ \hline 180 & 25 \end{array}$$

$$(180+(12+15)/2)(25)=>(193.5)(25)=19375$$

Solution: 125x155 = 19375

Problem 12.6: To find 125x145

$$\begin{array}{c|c} 12 & 5 \\ 14 & 5 \\ \hline 168 & 25 \end{array}$$

$$(168+(12+14)/2)(25)=>(181)(25) = 18125$$

Solution: 125x155 = 18125

EXERCISE PROBLEMS

The following are some of the exercise problems to be solved by using the method of Multiplication of numbers ending with the digit 5 to have an acquaintance with the new method of multiplication.

1.	25x85		16.	15x45
2.	45x65		17.	25x75
3.	45x85		18.	35x65
4.	35x75		19.	15x105
5.	35x95		20.	45x115
6.	25x105		21.	55x125
7.	45x125		22.	115x145
8.	45x165		23.	115x165
9.	125x125		24.	125x175
10.	75x135		25.	125x195
11.	75x75		26.	145x155
12.	85x85		27.	75x145
13.	25x125		28.	85x135
14.	35x115		29.	95x105
15.	135x155		30.	165x195

CHAPTER 13

SQUARING OF NUMBERS

In the preceding Chapters we have discussed the multiplication of two numbers based at 10, 50, 100, 500, 1000, 5000 and 10000 and is also illustrated with the help of several numerical examples. In the procedure of multiplication based at 10, one has to write one digit at a time; in the multiplication based at 100 one has to write two digits at a time whereas in the case of multiplication based at $5*10^k$ or 10^k one has to write k-digits at a time. Further we have also seen the multiplication of some special type of numbers and are explained with the help of several numerical examples. In this chapter we are planning to discuss some methods of finding the square of a given number. Firstly we are planning to discuss some methods of finding square of some specialized numbers and then we are planning to discuss a generalized method of finding a square of a number with any number of digits.

The other details of working procedures of these methods are explained in appropriate headings and also illustrated with the help of several numerical examples.

SQUARE OF A NUMBER ENDING WITH 5

The procedure of finding the square of a number ending with 5 is explained here. The method is also illustrated with the help of several numerical problems.

The steps involved are as follows:

i.   Let X_1 be the number to be squared.

ii.   Remove the rightmost digit 5 from the number and write the remainder, say R_1 at the left hand side and 5 at the right hand side.

iii.   Define the Remainder $R_2=R_1+1$ and write the remainder R_2 at the left hand side and 5 at the right hand side as done.

iv.   Find the product R_1*R_2.

v.   Write down the value R_1*R_2 below R_2 and 25 below the value 5 as done in Example 13.1.

vi.   Write down the digits from right to left to get the result.

For more details please refer the numerical examples given below:

Example 13.1: To find the square of 25, write down the numbers 25 and 35 vertically as written in the Figure 13.1. Find the product of 5 and 5; product of 2 and 3 and write down the results as done in the Figure 13.1. Write these numbers from right to left or left to right to get the square of 25 as 25x25=625

$$\begin{array}{c|c} 2 & 5 \\ 3 & 5 \\ \hline 6 & 25 \end{array}$$

$$(2*3)(5*5) => (6)(25)=625$$

Figure 13.1: 25x25 = 625

Example 13.2: To find the square of 45, write down the numbers 45 and 55 vertically as written in the Figure 13.2. Find the product of 5 and 5; product of 4 and 5 and write down the results as done in the Figure 13.2. Write these numbers from right to left or left to right to get the square of 45 as 45x45=2025

$$
\begin{array}{c|c}
4 & 5 \\
5 & 5 \\
\hline
20 & 25 \\
\end{array}
$$

$(4*5)(5*5) => (20)(25)=2025$

Figure 13.2: 45x45 = 2025

Example 13.3: To find the square of 55, write down the numbers 55 and 65 vertically as written in the Figure 13.3. Find the product of 5 and 5; product of 5 and 6 and write down the results as done in the Figure 13.3. Write these numbers from right to left or left to right to get the square of 55 as 55x55=3025

$$
\begin{array}{c|c}
5 & 5 \\
6 & 5 \\
\hline
30 & 25 \\
\end{array}
$$

$(5*6)(5*5) => (30)(25)=3025$

Figure 13.3: 55x55 = 3025

Example 13.4: To find the square of 75, write down the numbers 75 and 85 vertically as written in the Figure 13.4. Find the product of 5 and 5; product of 7 and 8 and write down the results as done in the Figure 13.4. Write these numbers from right to left or left to right to get the square of 75 as 75x75=5625

$$
\begin{array}{c|c}
7 & 5 \\
8 & 5 \\
\hline
56 & 25 \\
\end{array}
$$

$(7*8)(5*5) => (56)(25)=5625$

Figure 13.4: 75x75 = 5625

Example 13.5: To find the square of 105, write down the numbers 105 and 115 vertically as written in the Figure 13.5. Find the product of 5 and 5; product of 10 and 11 and write down the results as done in the Figure 13.5. Write these numbers from right to left or left to right to get the square of 105 as 105x105=11025

10	5
11	5
110	25

$$(10*5)(11*5) => (110)(25) = 11025$$

Figure 13.5: 105x105 = 11025

Example 13.6: To find the square of 155, write down the numbers 155 and 165 vertically as written in the Figure 13.6. Find the product of 5 and 5; product of 15 and 16 and write down the results as done in the Figure 13.6. Write these numbers from right to left or left to right to get the square of 155 as 155x155=24025

15	5
16	5
240	25

$$(15*16)(5*5) => (240)(25) = 24025$$

Figure 13.6: 155x155 = 24025

Example 13.7: To find the square of 395, write down the numbers 395 and 405 vertically as written in the Figure 13.7. Find the product of 5 and 5; product of 39 and 40 and write down the results as done in the Figure 13.7. Write these numbers from right to left or left to right to get the square of 395 as 395x395=156025

$$
\begin{array}{c|c}
39 & 5 \\
40 & 5 \\
\hline
1560 & 25
\end{array}
$$

$(39*40)(5*5)=> (1560)(25)=156025$

Figure 13.7: 395x395 = 156025

Example 13.8: To find the square of 995, write down the numbers 995 and 1005 vertically as written in the Figure 13.8. Find the product of 5 and 5; product of 99 and 100 and write down the results as done in the Figure 13.8. Write these numbers from right to left or left to right to get the square of 995 as 995x995=990025

$$
\begin{array}{c|c}
99 & 5 \\
100 & 5 \\
\hline
9900 & 25
\end{array}
$$

$(99*100)(5*5)=> (9900)(25)=990025$

Figure 13.8: 995x995 = 990025

SQUARING OF A NUMBERS NOT ENDING WITH 5

The procedure of finding the square of a number ending with other than 5 is explained here. Even though the proposed method is general in nature, but for the sake of convenience of the readers we discuss this method for the case of two digits. That is, the proposed method is explained for finding the square of two digit numbers. Later it may be explained for the general case. The method is also illustrated with the help of several numerical problems.

The steps involved in squaring of a two digit numbers are as follows:

i.   Let X is the two-digit number to be squared.

ii. Write the number X1 as (ab) where a is the left hand digit and b is the right hand digit.

iii. Find the squares of a, b and the product 2*a*b.

iv. Write down the values of a^2, b^2 and 2ab as done in the example 13.9.

v.   Write down one digit at a time from each bracket from right to left to get the result.

vi. If any value is more than 10 then reduce to mod 10 and carry away the remainders to the next brackets. Continue this step to write down all the digits.

For more details please refer the numerical examples given below:

Remark: While considering the square of n digit number, write the given n-digit number N as ab, where a is consists of first p digits of N and b is the last q digits of N such that n=p+q. Follow the same steps from i to iv as given in the above procedure except steps v and vi. While writing the digits from right to left write q digits at a time instead of one digit at a time. For more details please refer the examples 13.4 and 13.5.

Example 13.9: To find the square of 12, write down the number 12 as ab, where a=1 and b=2. Find the squares of 1; 2 and the product 2*1*2 and write down the results as done in the Figure 13.9. Write these numbers from right to left one digit at a time to get the square of 12 as 12x12=144

$$12^2 \quad = \quad (1^2)(2*1*2)(2^2)$$

$$12^2 \quad = \quad (1)(4)(4)$$

$$12^2 \quad = \quad 144$$

Figure 13.9: 12x12 = 144

Example 13.10: To find the square of 16, write down the number 16 as ab, where a=1 and b=6. Find the squares of 1; 6 and the product 2*1*6 and write down the results as done in the Figure 13.10. Write these numbers from right to left one digit at a time to get the square of 16 as 16x16=256

$$16^2 \quad = \quad (1^2)(2*1*6)(6^2)$$
$$16^2 \quad = \quad (1)(12)(36)$$
$$16^2 \quad = \quad (1)(12+3)(6)$$
$$16^2 \quad = \quad (1)(15)(6)$$
$$16^2 \quad = \quad (1+1)(5)(6)$$
$$16^2 \quad = \quad (2)(5)(6)$$
$$16^2 \quad = \quad 256$$

Figure 13.10: 16x16 = 256

Example 13.11: To find the square of 56, write down the number 56 as ab, where a=5 and b=6. Find the squares of 5; 6 and the product 2*5*6 and write down the results as done in the Figure 13.11. Write these numbers from right to left one digit at a time to get the square of 56 as 56x56=3136

$$56^2 \quad = \quad (5^2)(2*5*6)(6^2)$$
$$56^2 \quad = \quad (25)(60)(36)$$
$$56^2 \quad = \quad (25)(60+3)(6)$$
$$56^2 \quad = \quad (25)(63)(6)$$
$$56^2 \quad = \quad (25+6)(3)(6)$$
$$56^2 \quad = \quad (31)(3)(6)$$
$$56^2 \quad = \quad 3136$$

Figure 13.11: 56x56 = 3136

Example 13.12: To find the square of 84, write down the number 84 as ab, where a=8 and b=4. Find the squares of 8; 4 and the product 2*8*4 and write down the results as done in the Figure 13.12. Write these numbers from right to left one digit at a time to get the square of 84 as 84x84=7056

$$84^2 \quad = \quad (8^2)(2*8*4)(4^2)$$
$$84^2 \quad = \quad (64)(64)(16)$$
$$84^2 \quad = \quad (64)(64+1)(6)$$
$$84^2 \quad = \quad (64)(65)(6)$$
$$84^2 \quad = \quad (64+6)(5)(6)$$
$$84^2 \quad = \quad (70)(5)(6)$$
$$84^2 \quad = \quad 7056$$

Figure 13.12: 84x84 = 7056

Example 13.13: To find the square of 59, write down the number 59 as ab, where a=5 and b=9. Find the squares of 5; 9 and the product 2*5*9 and write down the results as done in the Figure 13.13. Write these numbers from right to left one digit at a time to get the square of 59 as 59x59=3481

$$59^2 \quad = \quad (5^2)(2*5*9)(9^2)$$
$$59^2 \quad = \quad (25)(90)(81)$$
$$59^2 \quad = \quad (25)(90+8)(1)$$
$$59^2 \quad = \quad (25)(98)(1)$$
$$59^2 \quad = \quad (25+9)(8)(1)$$
$$59^2 \quad = \quad (34)(8)(1)$$
$$59^2 \quad = \quad 3481$$

Figure 13.13: 59x59 = 3481

Example 13.14: To find the square of 512, write down the number 512 as ab, where a=5 and b=12. Find the squares of 5; 12 and the product 2*5*12 and write down the results as done in the Figure 13.14. Write these numbers from right to left two digits at a time to get the square of 512 as 512x512=262144

$$512^2 \quad = \quad (5^2)(2*5*12)(12^2)$$
$$512^2 \quad = \quad (25)(120)(144)$$
$$512^2 \quad = \quad (25)(120+1)(44)$$
$$512^2 \quad = \quad (25)(121)(44)$$
$$512^2 \quad = \quad (25+1)(21)(44)$$
$$512^2 \quad = \quad (26)(21)(44)$$
$$512^2 \quad = \quad 262144$$

Figure 13.12: 512x512 = 262144

Example 13.15: To find the square of 1512, write down the number 1512 as ab, where a=15 and b=12. Find the squares of 15; 12 and the product 2*15*12 and write down the results as done in the Figure 13.15. Write these numbers from right to left two digits at a time to get the square of 1512 as 1512x1512=2286144

$$1512^2 \quad =(15^2)(2*15*12)(12^2)$$
$$1512^2 \quad =(225)(360)(144)$$
$$1512^2 \quad =(225)(360+1)(44)$$
$$1512^2 \quad =(225)(361)(44)$$
$$1512^2 \quad =(225+3)(61)(44)$$
$$1512^2 \quad =(228)(61)(44)$$
$$1512^2 \quad =2286144$$

Figure 13.15: 1512x1512 = 2286144

SOLVED PROBLEMS

The following are some of the solved problems given to explain the method of squaring of numbers ending with 5.

Problem 13.1: To find 15x15

$$\begin{array}{c|c} 1 & 5 \\ 2 & 5 \\ \hline 2 & 25 \end{array}$$

(1*2)(5*5)=>(2)(25)=225

Solution: 15x15 = 225

Problem 13.2: To find 35x35

$$\begin{array}{c|c} 3 & 5 \\ 4 & 5 \\ \hline 12 & 25 \end{array}$$

(3*4)(5*5)=>(12)(25)=1225

Solution: 35x35 = 1225

Problem 13.3: To find 95x95

$$\begin{array}{c|c} 9 & 5 \\ 10 & 5 \\ \hline 90 & 25 \end{array}$$

(9*10)(5*5)=>(90)(25)=9025

Solution: 95x95 = 9025

Problem 13.4: To find 115x115

$$\begin{array}{c|c} 11 & 5 \\ \underline{12} & \underline{5} \\ 132 & 25 \end{array}$$

(11*12)(5*5)=>(132)(25)=13225

Solution: 115x115 = 13225

Problem 13.5: To find 135x135

$$\begin{array}{c|c} 13 & 5 \\ \underline{14} & \underline{5} \\ 182 & 25 \end{array}$$

(13*14)(5*5)=>(182)(25)=18225

Solution: 135x135 = 18225

Problem 13.6: To find 195x195

$$\begin{array}{c|c} 19 & 5 \\ 20 & 5 \\ \hline 380 & 25 \end{array}$$

(19*20)(5*5)=>(380)(25)=38025

Solution: 195x195 = 38025

The following are some of the solved problems given to explain the method of squaring of numbers not ending with 5.

Problem 13.7: To find 17x17. Let a=1 and b=7. Then

$$
\begin{array}{lcl}
17^2 & = & (1^2)(2*1*7)(7^2) \\
17^2 & = & (1)(14)(49) \\
17^2 & = & (1)(14+4)(9) \\
17^2 & = & (1)(18)(9) \\
17^2 & = & (1+1)(8)(9) \\
17^2 & = & (2)(8)(9) \\
17^2 & = & 289
\end{array}
$$

Solution: 17x17 = 289

Problem 13.8: To find 27x27. Let a=2 and b=7. Then

$$
\begin{aligned}
27^2 &= (2^2)(2*2*7)(7^2) \\
27^2 &= (4)(28)(49) \\
27^2 &= (4)(28+4)(9) \\
27^2 &= (4)(32)(9) \\
27^2 &= (4+3)(2)(9) \\
27^2 &= (7)(2)(9) \\
27^2 &= 729
\end{aligned}
$$

Solution: 27x27 = 729

Problem 13.9: To find 127x127. Let a=12 and b=7. Then

$$
\begin{aligned}
127^2 &= (12^2)(2*12*7)(7^2) \\
127^2 &= (144)(168)(49) \\
127^2 &= (144)(168+4)(9) \\
127^2 &= (144)(172)(9) \\
127^2 &= (144+17)(2)(9) \\
127^2 &= (161)(2)(9) \\
127^2 &= 16129
\end{aligned}
$$

Solution: 127x127 = 16129

Problem 13.10: To find 121x121. Let a=12 and b=1. Then

$$
\begin{aligned}
121^2 &= (12^2)(2*12*1)(1^2) \\
121^2 &= (144)(24)(1) \\
121^2 &= (144+2)(4)(1) \\
121^2 &= (146)(4)(1)
\end{aligned}
$$

Solution: 121x121 = 14641

Problem 13.11: To find 121x121. Let a=1 and b=21. Then

$$
\begin{aligned}
121^2 &= (1^2)(2*1*21)(21^2) \\
121^2 &= (1)(42)(441) \\
121^2 &= (1)(42+4)(41) \\
121^2 &= (1)(46)(41)
\end{aligned}
$$

Solution: 121x121 = 14641

EXERCISE PROBLEMS

The following are some of the exercise problems to be solved by using the method of squaring of numbers ending with the digit 5 and not ending with 5 to have an acquaintance with the new methods of squaring.

1.	65x65		16.	18x18
2.	45x45		17.	26x26
3.	85x85		18.	32x32
4.	315x315		19.	83x83
5.	305x305		20.	94x94
6.	225x225		21.	152x152
7.	245x245		22.	119x119
8.	415x415		23.	211x211
9.	525x525		24.	123x123
10.	175x175		25.	222x222
11.	715x715		26.	1112x1112
12.	185x185		27.	1313x1313
13.	255x255		28.	1215x1115
14.	435x435		29.	2121x2121
15.	235x235		30.	1616x1616

CHAPTER 14

MATHEMATICAL PUZZLES – SIMPLE DIAGRAMS AND NUMERICAL PROBLEMS

As discussed in Chapter I, in day-to-day operations everyone in the society has to do some sort of mathematical computations due to the nature of their works or businesses or services. In fact it is necessary for people from school students to research scientists and ordinary laymen to top level businessmen for the purpose of computing the school exercises; establishing research findings; calculating their earnings or expenses; assessing their turn over or marketing share, etc. It is a common feeling that almost every one in the world has some skills on Mathematics and Statistics and may take some efforts to acquire them. On the contrary as pointed out by Keith Devlin of Stanford University, USA "The real world is one in which there is math phobia among a sizeable minority, some level of math anxiety among many more, a general antipathy towards mathematics in the majority and ignorance about the true nature of mathematics on the part of practically everyone but the professional mathematician". What is required to popularize the mathematical and statistical thinking among the people particularly among the students community? By making mathematics and statistics as a subject to be taught in much the same way as History or Geography or English literature; it is to be taught, not as a utilitarian toolbox but as a part of human culture. Consequently the teaching of mathematics and statistics may create an awareness of the nature of these subjects and their role in contemporary society.

To eliminate the math phobia and to create interests in the subject of mathematics and statistics, one may plan to deliver the subject matter through Recreational Mathematics. It is a common phenomenon that every human being has to do some recreational activities to keep the mind free from stress and tension. The recreation means that the activities that we do when we are not working. There are a number of ways one can plan for recreation. For example visiting picnic spots/ cinema/ friends/ relatives/ clubs, playing games/ gambling, watching television/ video, doing social and community services, conducting physical and chemical experiments, reading books/newspapers/novels, doing stamp collection, recreational mathematics, etc. are some of the activities one can think of for recreations. Now the question is what makes the difference among the several recreational activities mentioned above? Naturally the answer is the activity, which requires less money; no constraints on the number of people involved and the nature of end results. By considering these points the recreational mathematics is preferable compared to other recreational activities due to the following: It requires only papers and pencils; no constraints of people, place, time and other factors. Then there is no tragedy in the end results unlike in visiting picnic spots; the visits may end with tragedy due to the road accidents. It is very much evident that there are number of road accidents taking place in every part of our country.

In recreational mathematics one can do some mathematical activities which include: Modelling (diagrams and graphs), Numbers (Games and patterns), Magic squares, Magic rectangles, creating and solving mathematical puzzles, reading the history of mathematics and mathematicians, search for new applications, etc.

What are available in magic squares? In magic squares one can look for the existence criteria, methods of construction, patterns among magic squares, number of possible solutions, relationships with other scientific areas, applications in real life problems, developing software, etc. It has been proved that there is no magic square of order 2. By rotation and reflection one may obtain 8, 7040 and 275305224 magic squares of order 3, 4 and 5 respectively. For other orders of magic squares the number of possible solutions available is not known. In looking at the applications of magic squares it is found that magic squares are very much useful for the selection of a systematic sample and in planning experimental designs in biological applications.

If we look at the advantages of the recreational mathematics one may conclude the following: Recreational mathematics eliminates math phobia, creates interests in mathematics, stimulates learning, improves memory/ knowledge/ numerical ability, develops manipulative skills, creates lateral thinking/ statistical thinking, and improves self satisfaction/ powers of reasoning. It is cost effective, it has no constraints of time/ place/ people, improves decision-making process and safety aspects.

As a concluding remark one can think of recreational mathematics whenever planning for recreational activities, which have a lot of advantages and those who find time think of solving the mathematical puzzle given below. It will certainly enhance your analytical and logical skills.

Puzzle 14.1: Count the number of triangles to be formed from the figures given below:

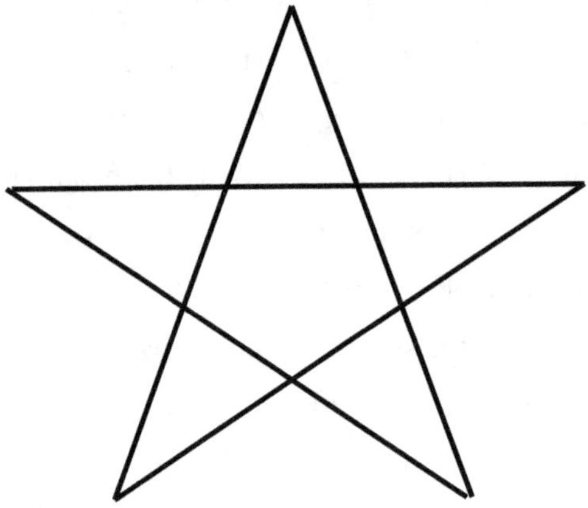

Figure 14.1

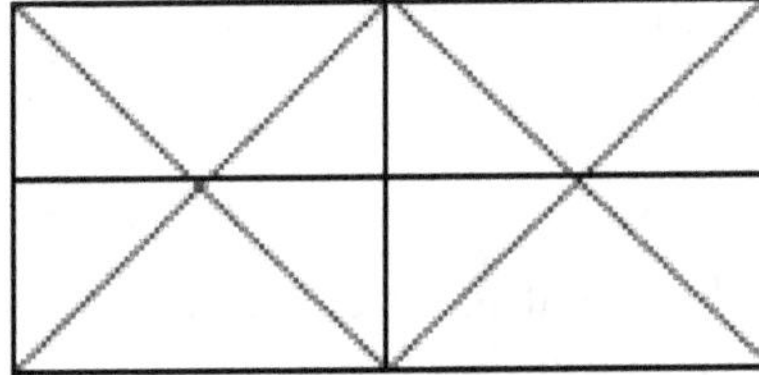

Figure 14.2

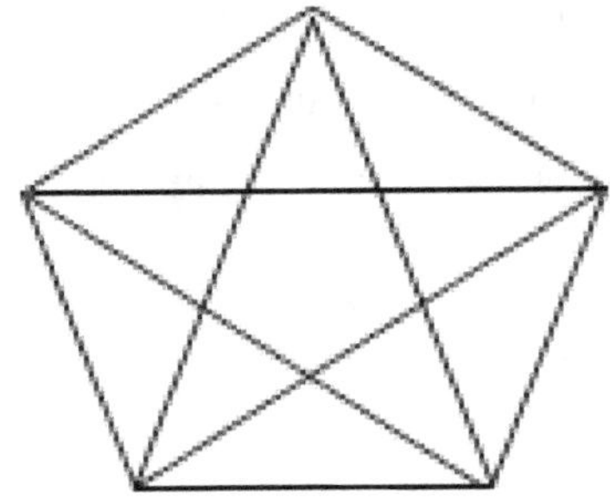

Figure 14.3

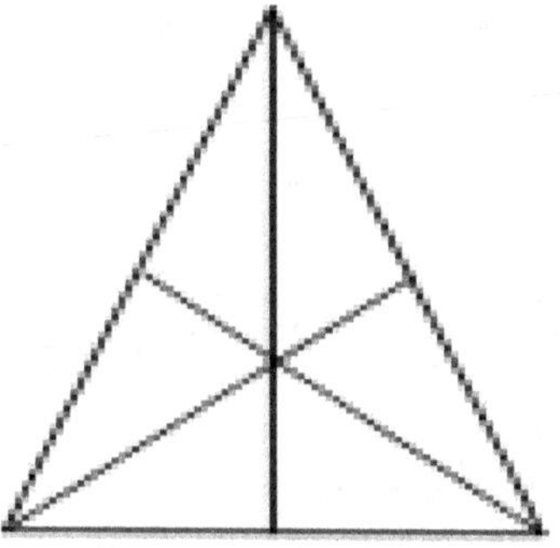

Figure 14.4

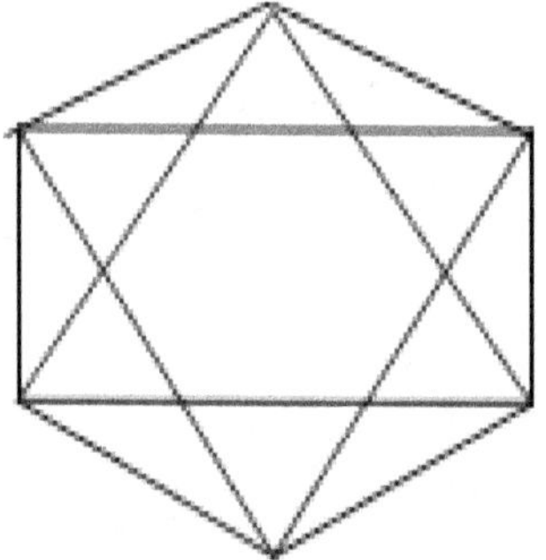

Figure 14.5

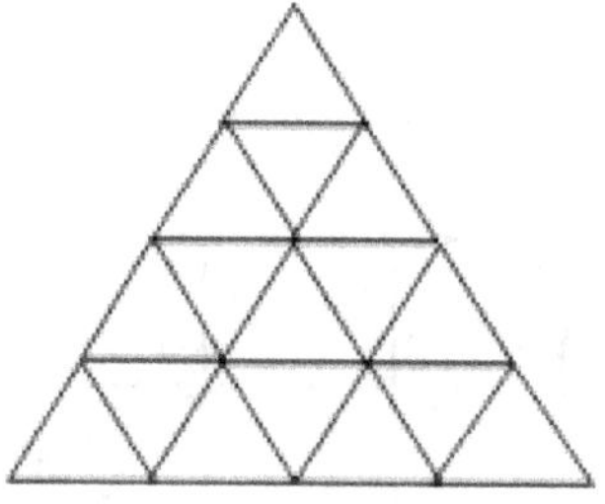

Figure 14.6

Puzzle 14.2. Can you draw the following figures by a continuous line without crossing any line?

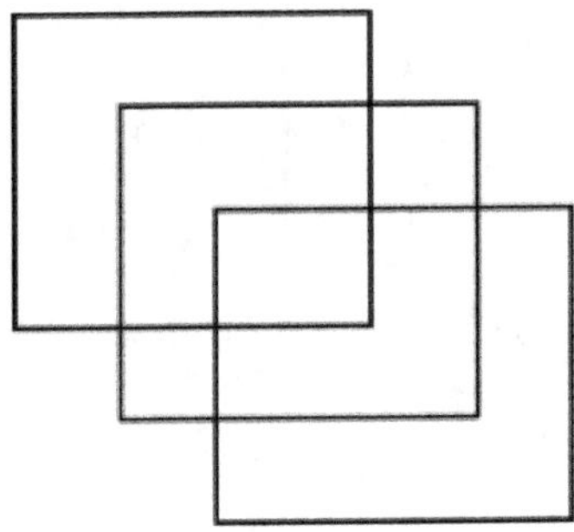

Figure 14.7

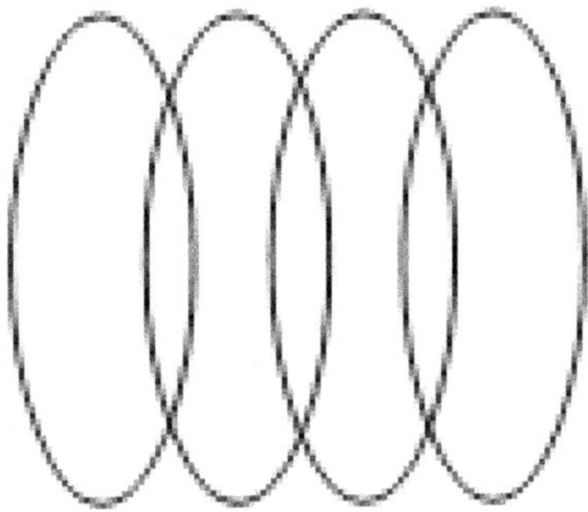

Figure 14.8

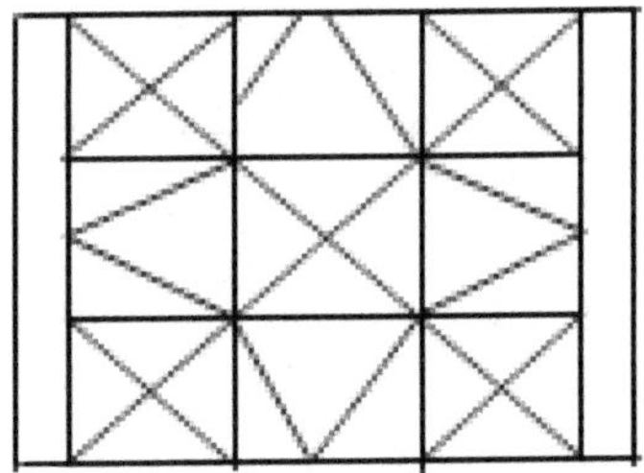

Figure 14.9

Puzzle 14.3: Can you split the following figure into 2, 3, 4, 6, 8, and 9 equivalent parts?

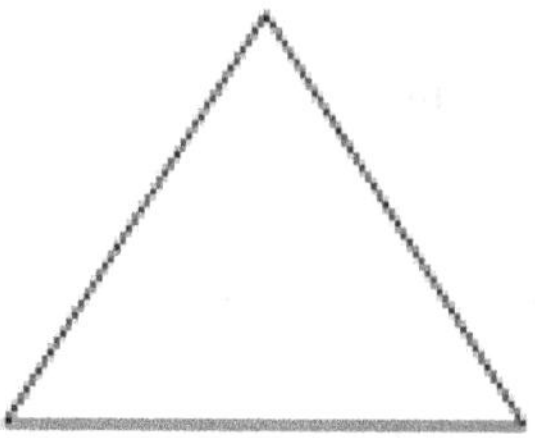

Figure 14.10

Puzzle 14.4: can you split the following figures into four equivalent parts?

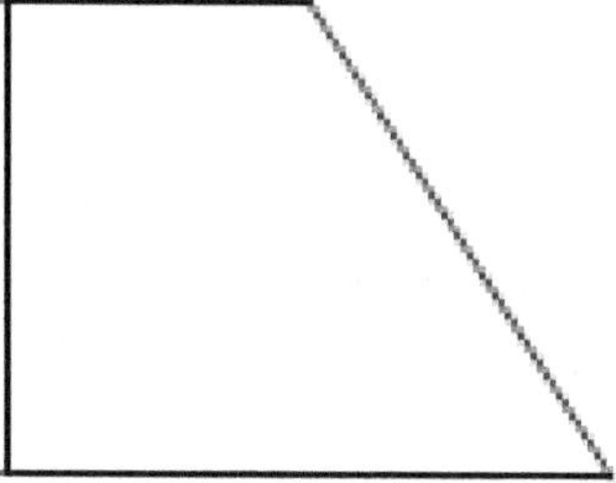

Figure 14.11

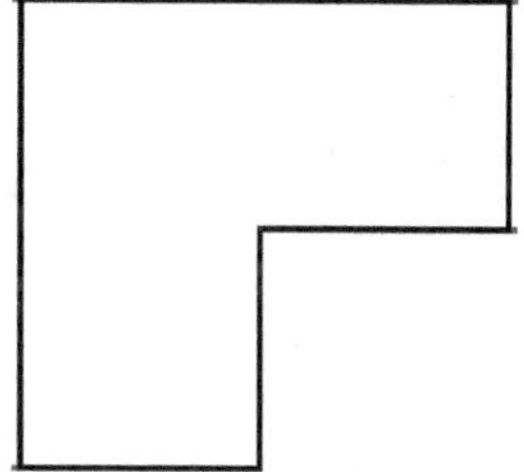

Figure 14.12

Puzzle 14.5. If sum of the two numbers is 100 and the difference is 60 then find the numbers.

Puzzle 14.6. If the cost of 5 apples is same as the cost of 3 apples and 4 oranges then find the ratio of the cost of one apple and one orange.

Puzzle 14.7. Six men and 9 women can do a piece of work in 10 days. How many days 2 men and 3 women will take to complete the work?

Puzzle 14.8. In 40 liters of mixture of milk and water, the quantity of water is 20%. How much water should be added to make the quantity of water and milk is equal?

Puzzle 14.9. A bag contains 50 coins in the denomination of 1, 5 and 10 paisa coins. If the total value of the coins is 100 paisa then find the number of coins in each of the denomination.

Puzzle 14.10. If Chitra is 7 years older than her younger sister Sivagami and after five years from now her age becomes twice of Sivagami then find the ages of Chitra and Sivagami.

Puzzle 14.11. If a boy took as much as time in running 10 meters, as his father took in covering 25 meters then find the distance covered by the boy during the time his father covered 500 meters.

Puzzle 14.12. If Karthika took 40 minutes to reach the school by cycling at the speed of 12 kms per hour then determine the speed required to reach the school 10 minutes earlier.

www.ingramcontent.com/pod-product-compliance
Lightning Source LLC
LaVergne TN
LVHW041318200726
843509LV00009B/540

9 789385 477706